Lecture Notes in Computer Sc

Commenced Publication in 1973
Founding and Former Series Editors:
Gerhard Goos, Juris Hartmanis, and Jan van Leeuwﻪ

Emanuele Salerno   A. Enis Çetin
Ovidio Salvetti (Eds.)

Computational Intelligence for Multimedia Understanding

International Workshop, MUSCLE 2011
Pisa, Italy, December 13-15, 2011
Revised Selected Papers

 Springer

Volume Editors

Emanuele Salerno
Ovidio Salvetti
ISTI
National Research Council of Italy
Via Moruzzi, 1
56124 Pisa, Italy
E-mail:{emanuele.salerno, ovidio.salvetti}@isti.cnr.it

A. Enis Çetin
Bilkent University
Electrical and Electronics Engineering
06800 Bilkent, Ankara, Turkey
E-mail: cetin@bilkent.edu

ISSN 0302-9743   e-ISSN 1611-3349
ISBN 978-3-642-32435-2   e-ISBN 978-3-642-32436-9
DOI 10.1007/978-3-642-32436-9
Springer Heidelberg Dordrecht London New York

Library of Congress Control Number: 2012946551

CR Subject Classification (1998): H.5.1, I.4, I.2.10, I.5, I.2.4, I.2.6, H.2.8, G.1.2

LNCS Sublibrary: SL 6 – Image Processing, Computer Vision, Pattern Recognition,
and Graphics

Typesetting: Camera-ready by author, data conversion by Scientific Publishing Services, Chennai, India

Printed on acid-free paper

Springer is part of Springer Science+Business Media (www.springer.com)

Preface

We live in a world where multimedia technology and resources improve the quality of our daily experiences. Multimedia data, in the form of still pictures, graphics, 3D models, audio, speech, video and a combination of these are in fact playing an increasingly important role in our life. Multimedia offers awe-inspiring possibilities to meet modern needs by enabling the distribution of content through several means. This information can be processed to create new knowledge. Therefore, we need to develop algorithms for computational interpretation and processing of such data, but also to share them across the network in an intelligent manner. Networked multimedia communications, such as video and audio on demand, video conferencing and e-learning, give us an idea of the growing diffusion of these data. Multimedia communications produce a growing demand for systems and tools to satisfy the most sophisticated requirements for storing, searching, accessing, retrieving, managing and sharing complex resources with many different formats of media types. A multimedia system is generally made of different components: database, multimedia storage server, network and client systems, often in a mobile environment. New initiatives are trying to bind these components together in open multimedia frameworks for an interoperable use of multimedia data in a distributed environment. Finding multimedia objects by content in a distributed database means searching for them on the basis of content descriptions and similarity measures. Looking for events in a video library requires not only searching a database for objects related to the specified events, but also considering bandwidth constraints. Enabling automatic classification and segmentation of digital resources is necessary to produce information that can be then extracted from multimedia also considering annotations. The above processes involve multimedia producers, but also several scientific communities (*e.g.*, cultural heritage, education, medicine, etc.) who want to merge multimedia description, management, and processing for their domain-specific requirements. This entails a proper understanding of the semantic relationships between the multimedia requisites in different domains. Research in multimedia analysis and understanding ranges from low-level data processing to automatic recognition. Current research domains involve methods and techniques for pattern recognition, content-based indexing, image understanding, object/subject-based retrieval and representation, segmentation, multimodal and multisource signal fusion, as well as knowledge processing and tracking. Multimedia understanding has a wide range of applications. Fundamental research and new technologies give rise to the growth of new understanding and applications outcomes.

March 2012

A. Enis Çetin
Ovidio Salvetti

Introduction

This book contains the contributions presented at the International Workshop on Computational Intelligence for Multimedia Understanding, organized by the ERCIM working group on Multimedia Understanding Through Semantics, Computation, and Learning (MUSCLE) as an open forum to discuss the most recent advances in this field. The workshop took place in Pisa, Italy, during December 13–15, 2011.

An open call-for-papers was issued in May 2011. After a first review, the contributions that scored more than 4/5 (43% of the original submissions) were accepted for publication. The papers that scored no more than 3/5 were rejected. Among the other papers, those that scored 4/5 were accepted, but their second versions were subjected to scrutiny by the editors, and the ones that scored between 3/5 and 4/5 were given the possibility to be presented at the workshop, but their second versions underwent a new full review before final acceptance. In this way, with the help of more than 20 external reviewers, the Program Committee managed to ensure both a high scientific level and an acceptance rate of more than 80%. I am highly indebted to the members of the Program Committee and the reviewers for their hard work. I must also thank all the contributors, whose presentations made the workshop extremely interesting and fruitful, and whose authoritative papers give value to this book. Two greatly appreciated invited lectures were given by Sanni Siltanen of VTT, Finland, and Bülent Sankur of Bogaziçi University, Turkey. I thank them very much on behalf of the entire audience.

The papers presented here are both theoretical and application-oriented, covering a wide range of subjects. As far as the media are concerned, most contributions deal with still images, but there are some on video and text. Two papers also use results from human perception experiments for image and video analysis and synthesis. The organization of the volume follows, where possible, both criteria of media/methodology and application. The first three papers are all related to meaning and semantics. The papers by Perner and Colantonio et al. are devoted to ontologies, a very important subject for knowledge-based media analysis, also devising the possibility of including the processing algorithms in the ontological reference. Then, the only paper dealing with text analysis, by Malandrakis et al., describes a technique to detect the affective meaning of text fragments to expand specialized lexica. The papers by Ruiz et al., Kayabol et al., and Szirányi and Szolgay deal with classification, with different applications and approaches. Classification also emerges as one of the most important issues in media understanding, since, with segmentation, it is probably an essential prerequisite for the extraction of meaning. Another often essential step to extraction of meaning is shape reconstruction. The papers by Kim and Dahyot and Kovács are devoted to this topic. Keskin and Çetin improve on fractional wavelets to build a powerful signal analysis tool. Han et al. treat spatial registration, another essential task when dealing with multimodal data. They adopt a geometric approach, pointing out that raw data or low-level features may not be appropriate to match images that are different in nature. Four papers from the same research team (Vácha and Haindl, Haindl and V. Havlíček, Haindl and M. Havlíček, and Filip et al.) deal with texture modeling, synthesis, and recognition, which are relevant to multimedia understanding and visualization, and also affect the fixation/attention of human observers. Carcel et al., Martienne et al., and Gallego et al. deal with the video media, describing annotation, labeling, and segmentation strategies, respectively. Magrini et al. present a real-time image analysis system deployed in an infomobility scenario. Finally, Szalai et al. describe a statistical model of human observers' behavior that helps to estimate possible tracking paths and regions of interest of the human visual system presented with video data.

Just to add a general remark to this conceptual map, I would say that there is a ubiquitous quest in this volume, which is mentioned explicitly in the Preface: "information can be processed to create new knowledge." This is not immediate, and trying to work the details out would drive us to ancient philosophical questions. As often happens, great poetry helps us:[1]

> *Where is the wisdom we have lost in knowledge?*
> *Where is the knowledge we have lost in information?*

Well, these papers show us how information processing can bring us to *meaning*. Perhaps meaning processing could bring us nearer to *knowledge*. Also, knowledge processing could bring us to wisdom, but this is out of the scope of this volume.

March 2012   Emanuele Salerno

---

[1] T. S. Eliot, *The Rock*, 1934

Organization

MUSCLE 2011 was organized by ERCIM, the European Research Consortium in Informatics and Mathematics, Working Group on Multimedia Understanding through Semantics, Computation and Learning.

Workshop Chairs

General Chairs

A. Enis Çetin   Bilkent University, Turkey
Ovidio Salvetti   National Research Council of Italy

Program Chair

Emanuele Salerno   National Research Council of Italy

Program Committee

Rozenn Dahyot   Trinity College Dublin, Ireland
Patrick Gros   INRIA, France
Michal Haindl   UTIA, Czech Republic
Nahum Kiryati   Tel Aviv University, Israel
Ercan E. Kuruoğlu   CNR, Italy
Marie-Colette van Lieshout   CWI, The Netherlands
Rafael Molina   University of Granada, Spain
Montse Pardás   Universitat Politècnica de Catalunya, Spain
Petra Perner   IBAI, Germany
Béatrice Pesquet-Popescu   Telecom Paristech, France
Ioannis Pitas   Aristotle University of Tessaloniki, Greece
Alex Potamianos   Technical University of Crete, Greece
Sanni Siltanen   VTT, Finland
Tamas Sziranyi   SZTAKI, Hungary
Anna Tonazzini   CNR, Italy
Simon Wilson   Trinity College Dublin, Ireland
Josiane Zerubia   INRIA, France

Local Committee

Francesca Martinelli   CNR, Italy
Davide Moroni   CNR, Italy, Chair
Gabriele Pieri   CNR, Italy
Marco Tampucci   CNR, Italy
Ettore Ricciardi   COCES and ISTI-CNR, Italy, Secretariat

Referees

G. Amato	P. Gros	E. Salerno
M. Belkhatir	M. Haindl	O. Salvetti
G. Bianco	L. Havasi	P. Savino
A. E. Çetin	K. Kilic	F. Sebastiani
G. Charpiat	N. Kiryati	J. Shi
S. Colantonio	R. Molina	S. Siltanen
M. D'Acunto	D. Moroni	C. Strapparava
L. Dempere	S. Pankhanti	T. Sziranyi
R. Dahyot	M. Pardás	A. Tonazzini
J. Estévez	G. Pieri	A. Utasi
G. Franzè	I. Pitas	D. Vitulano
C. Gennaro	P. Perner	S. Wilson
C. Germain	A. Potamianos	B. Zitová

Sponsoring Institutions

European Research Consortium in Informatics and Mathematics - ERCIM

National Research Council - CNR, Italy

National Institute for Research in Informatics and Automation - INRIA, France

Table of Contents

Learning an Ontology for Visual Tasks

Petra Perner

Institute of Computer Vision and Applied Computer Sciences, IBaI
Kohlenstr. 2, Leipzig, Germany
pperner@ibai-institut.de

Abstract. The definition of ontology for visual tasks is often very tricky, since humans are usually not so good at expressing visual knowledge. There is a gap between showing and naming. The knowledge of expressing visual experience is often not trained. Therefore, a methodology is needed of how to acquire and express visual knowledge. This methodology should become a standard for visual tasks, independent of the technical or medical discipline. In this paper we describe the problems with visual knowledge acquisition and discuss corresponding techniques. For visual classification tasks, such as a technical defect classification or a medical object classification, we propose a tool based on the repertory grid method and image-processing methods that can teach a human the vocabulary and the relationship between the objects. This knowledge will form the ontology for a visual inspection task.

1 Introduction

With the introduction of digital image-acquisition units in many technical, medical or personal applications, the need for describing and processing visual content increased. In long-lasting applications such as medical radiography, medical doctors developed based on their experience an ontology called the BIRAD´s code [1] that helps to describe the content of medical radiological images in e.g breast-cancer images and pulmonary images. This code is used for training newcomers in the field and for the development of automatic image-interpretation systems [2].

In many defect-inspection tasks, such as wafer inspection, the technical engineers developed over time a vocabulary and rules for the description of the defects [3]. The development of this vocabulary was a long process driven by the need to have a generalized vocabulary and rules to train the operator on the lines. The vocabulary is based on the personal impressions of the engineers. The terms are usually very specific terms that might not be standard expression terms for a certain visual appearance.

Early work on knowledge acquisition in visual interpretation tasks was reported by Hedengren [4], Lundsteen et al. [5], Szafarska et al. [6], and Perner [7].

In both of the above described applications no automatic tool supported the process, nor was there a methodology behind that helped to carry out the process in an efficient way.

E. Salerno, A.E. Çetin, and O. Salvetti (Eds.): MUSCLE 2011, LNCS 7252, pp. 1–16, 2012.
© Springer-Verlag Berlin Heidelberg 2012

Visual tasks are very different and therefore need different automatic methods for learning the ontology. There are image-retrieval tasks to be done, image-inspection tasks, and image-interpretation tasks of complex images or videos. A classification of these tasks should help to come up with a standard for knowledge acquisition.

Besides that a visual term is first of all a symbolic description. Image-processing procedures should be able to acquire the symbolic description for an object or part of an object from the image. That means that a visual ontology is always comprised of two tasks: first, a standard visual description term should be defined that describes the visual content well and its meaning should easily be understood by different operators/users, and second, the mapping of the symbolic visual description term to low or high-level image descriptions given by image processing methods should be possible.

In this paper we describe the problem of learning an ontology based on a visual inspection task. The right visual terms are not known from scratch. It is only while applying the repertory grid method [8] that these visual terms and their relationship to each other will be learnt. At the same time, the similarity relationship between the different objects will be learnt. Only visual terms that make sense and can differentate between different objects will be chosen by the method. We intend to use this task procedure for technical domains, but consider our method also applicable to medical or other applications. In general we think that this can be a standard technique for image classification of objects. As for us learning an ontology has to do with knowledge acquisition, we also consider it as knowledge acquisition. The efforts to learn a visual ontology should lead to more standard visual description terms and image-processing procedures that can acquire this knowledge from images.

In section 2 we describe the main problem of visual inspection. The knowledge-acquisition problem is explained in section 3. An overview of knowledge-acquisition problems is given in section 4. In section 5 we describe the repertory grid method. The application of the repertory grid method to a visual inspection task is described in section 6. Finally, we describe the architecture of our knowledge-acquisition tool based on the repertory grid method and image-processing methods in section 7. Conclusions are given in section 8.

2   The Visual Inspection Task

Any human thought can be expressed as a proposition and the formulation of a proposition is hardly possible without the help of "concepts", which are usually supposed to correspond to certain classes of objects or situations. A sentence expressing the perception "I see a light object" shows that there is a concept "intensity".

In this section, we want to describe a model for the inspection task, in order to discover the main concepts which underlay the specific domain.

The performance of the task is shown in figure 1. The operator has to inspect an industrial part for defects. An image-acquisition system is involved in most of the inspection tasks, e.g. a microscope, an x-ray inspection system or an ultrasonic imaging system.

The inspected object generally consists of the following parts:

- various 2-d or 3-d geometrical structures
- various material components, like glass or metal
- relationships between the structures.

With this information an operator forms an object model for the defect of the inspected industrial part.

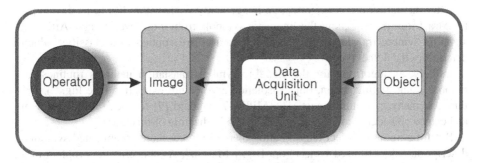

Fig. 1. Inspection Task

The defect itself might have the following properties:

- absolute attributes, like roundness, fuzzy margin, light area and so on
- relative attributes, which are defined by comparing the defect to other objects in the object model, and
- relationships to other objects in the object model, like "the defect is located on object O_1".

The behavior of the image-acquisition system plays an important role in the inspection, because it determines how a certain object is transformed into an image. The following example shows how a an operator uses his own mental model of the image-acquisition-system behavior to make his decision:

A microscope is focused at layer_2, which should be the surface of the original object. The defect object lies on layer_1 which is above layer_2. The defect object is out of focus and will be seen in an image as a blurred object.

Seeing a blurred object, the operator concludes that there is something on top of the original object.

Models of the inspected object, the defect and the image-acquisition system combine to form the conceptual model of a human visual inspection behavior. The behavior of the inspection system puts some constraints on the appearance of any defect. This in turn constrains the attributes and the geometric description.

To make this conceptual model explicit, it helps to better understand the process of defect classification.

The formalism will provide a common language for expressing the components of the model and the appearances of the defects.

3   The Knowledge-Acquisition Problem

The basis for the development of an automatic visual defect-classification system is usually the human's linguistic description about the appearance of the defect types. In most applications, this description is available, since humans have been doing inspections over a long time and have developed substantial knowledge about that. Nowadays, applications are different. The knowledge about the visual description of an object has to be developed in a short time.

After the definition of the terms, an image-analysis algorithm has to be developed to extract information about the linguistic symbols from a digitized image. After the signal-to-symbol translation the symbolic features are applied to a classifier which determines the defect type.

Usually, an algorithm for the signal-to-symbol translation includes routines for preprocessing, segmentation, object isolation and feature extraction (see Fig. 2). For this task many algorithms exist. The MPEG-7 standard [9] focuses on this problem and can be considered as a first step into a standardization of this task.

But for automatic defect classification there are many more complicated symbolic descriptions like "Blurred Object" or "Fuzzy Margin". As concerns the problem of a human to fit these symbolic descriptions into a machine-understandable form, new image-processing procedures have to be developed. Usually, every new application will require a new symbolic description and a few new image-processing procedures to extract this information from the image. On the other hand, to have a standard collection of visual terms and to know the meaning of these terms will help to standardize visual description tasks. The vocabulary will not explode by particular terms defined by a human. It will also lead to a better ability of humans to describe visual contents.

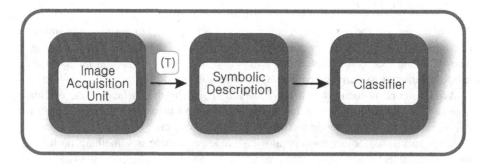

Fig. 2. Defect Classification System

A first step in that direction has been done by Rao [10] for texture description. Systems following that idea have been proposed by Trusova et al. [11] and Perner [12] for medical purposes and by Clouard et al. [13].

Defining ontology for visual tasks will cause the number of image processing algorithms for mapping the symbolic terms into machine-readable terms to grow continuously. The problem of getting a description of visual content from a human is related to the knowledge-engineering problem and will result into a visual ontology for the particular application. We will describe this process in the next section.

4 Recent Knowledge-Acquisition Techniques

There are several knowledge-acquisition techniques [14] known from psychology and knowledge-engineering literature. These techniques range from interview techniques, to methods for dichotomization, to card sorting for the classification of objects, to classification methods, to methods for similarity determination like clustering, and to multidimensional scaling. We describe the methods that are interesting for visual content acquisition.

4.1 The Interview Technique

A widely used technique for knowledge acquisition is the interview. First, an expert is asked to give a description of a defect image:
"Please describe the defect image."
"How many dark, vertical lines exist in the light region of the image?"
The preliminary step is to conceptually segment or partition the image into one or more regions as was done by the natural language description?.
Each region may contain structures of interest and interfering background. Of course, what is called 'structure' and what is called 'interfering background' depends on the application [4].
Rather than a description, for the classification problem a questionnaire is necessary to acquire distinguishing features. The question type for distinguishing features can be:
"What property distinguishes one defect from another?"
When a knowledge engineer has obtained the description and the distinguishing features, he has to determine if the knowledge is complete and consistent. A good way to do that for defect classification is to create a decision table, where the rows are the defect types and the columns the attributes. Every blank field in the decision table shows a knowledge gap. From the decision table, it is possible to form a decision tree for the classification [7]. The decision tree is based on the expert description. For an automatic inspection system, a signal-to-symbol transformation has to be done after that.
The description of an image is easy, if the person can use or has in his mind a well-defined vocabulary for the description of the image content. This implies the information about the physical properties of objects or about physical relationships among objects explicitly encoded at some previous time and that any human expert understands the visual concept behind these descriptions. This is often not the case. Humans tend to draw a picture of a defect rather than describe the defect in natural language. The main reason for that is that humans do not have a good visual vocabulary and its visual meaning in their minds. The work on ontologies will result in a better vocabulary and understanding of the visual meaning of these terms.

Usually, during the knowledge-engineering process in several steps the implicit knowledge about the defect is made explicit in words.

As shown in [15], a drawn picture cannot be a mental photograph of a defect, but it will have some visual characteristics of the real object. Those characteristics can be useful in retrieving information about the visual appearance of objects. The question is what we can use from this pictorial information for the problem of defect classification.

4.2   The Interactive Image-Analysis System

Interactive image-analysis systems are widely used in medical domains like computer tomography or cell analysis [11]-[13]. They are also known in industrial inspection systems [4]. Such systems provide some applicable image processing, image-analysis and feature-extraction algorithms and also interactive graphical facilities based on light pens or boards. A user can specify the region of interest and apply certain procedures to the specified region. The results from the image analysis are marked in the original image. Every image-analysis algorithm stands for a symbolic description of the image properties. By applying these procedures to the image and observing the result, the user gets a confirmation of the meaning of his description. If the result does not match his meaning, he can think of a better description of the image property.

Such a system requires an initial description of the problem domain, which serves as a basis for the development of applicable image-analysis procedures. This set of image-analysis procedures will grow with every newly discovered attribute for the property of the image in that specific domain. On the basis of the discovered attributes one can form an ontology for the description of the image. A natural language programming system is described in [5].

4.3   The Image Catalogue

Because it is time-consuming and difficult to describe all the properties of defect images, people tend to collect defect catalogues rather than describe the properties of the defect completely [6][7]. In this catalogue they store an image for every defect type which is supposed to be a new defect. In addition to the images the catalogue contains a short description of the defect and information about the technological process.

People use this catalogue for operator training or to assist an operator when the decision is not easy to make besser: is difficult. He would look up the catalogue, select the images having the same technological information and try to make a decision about the defect by comparing the real defect image with the collected defect images in the catalogue. If he finds that one defect image in the catalogue is similar to the real image, the real defect gets the name of the defect image. This approach addresses the problem of the silent feature. What cannot be described easily by words is contained silently in the image.

4.4   Motivation for a Knowledge-Acquisition Tool Based on the Repertory Grid Method

As described in Sect. 3, the goal of defect classification is to ask for distinguishing features. A specific questionnaire asking for distinguishing features is necessary. This questionnaire should help to minimize the influence of human bias in the description. An automatic knowledge-engineering tool should be based on such a questionaire and be able to evaluate the quality of the entered knowledge. The tool should provide automatically generated inspection rules on the basis of the elicited knowledge and should update the knowledge and the rules with every new observed defect. Furthermore, such a system should be developed based on the need of a human inspector. A tool based on the repertory grid method [16] seems to satisfy many of these requirements. This method provides a specific questionnaire which seems to be appropriate for the classification task. The method allows for an evaluation of the knowledge on the basis of cluster analysis or multidimensional scaling methods. Boose in particular has reported considerable success in the computer automation of the knowledge-extraction process [17].

5   The Repertory Grid Method

Free listing of features has often been used by cognitive psychologists studying the nature of psychological concepts and categories. Certain problems have been noted with the lists obtained by this method. For example, the set of features listed may be biased by the frequency of the terms used to describe the features, continuous or dimensional features may be difficult to describe, verbal associates rather than conceptual aspects of the objects may be listed, and idiosyncratic responses may be given.

Kelly developed a theory of human cognition on the primitive notion of constructs, or dichotomous distinctions, which are "templates that a person creates and attempts to fit over the realities of which the world is composed" [15]. He proposed that all of human activity can be seen as a process of anticipating the future by construing the replication of events, and revising constructs according to the success of this anticipation. Kelly developed his theory in the context of clinical psychology and hence was concerned with techniques used to by-pass cognitive defenses and elicit the construct systems underlying behavior. In a non-clinical context this is the problem of knowledge engineering discussed above.

In the repertory grid method, features are elicited by asking the subjects to compare the objects in the set of choice alternatives in groups of three. When groups of three objects are used, the subject is asked to name a feature that two of them have in common and that distinguish them from the third. The common feature and the opposite feature form a construct. The user is asked to rate each object on a scale from 1 to 5, considering the features. Such a scale is shown with the two features "longelongated" and "circular" which belong to the concept "shape" in Fig. 3. After defining constructs and rating each element, the program evaluates the resulting matrix using cluster analysis techniques [17]. The degree of closeness among elements, based on the chosen constructs, is then displayed.

When elements are closely related, some new constructs that can better distinguish among the objects should be found. If no such construct is found, then the objects are deemed redundant, and one or more objects should be combined. Similarly, the program suggests constructs that can be clustered and deleted if appropriate.

Fig. 4 shows the result from a repertory grid process. The dendrogram was constructed by the single linkage method [18]. The first two constructs "longelongated" and "random shape" are 100% similar and can be merged together. This means that the subject was able to give this description for the shape, but in principle the subject could not clearly assign one of these two features for the shapes to the objects. These features do not have any meaning for the specific application, or so there should be given by the subjects the operators a more detailed or precise description of the shape.

When we are satisfied with the results, we want to have rules for the classification based on the elicited knowledge. One way to do this is to ask the expert to label the dendrogram as suggested by Butler et al. [16]. This procedure works well for little decision trees and crisp data.

Gaines and Shaw [19] developed an algorithm based on fuzzy logic, which gives the entailment between the constructs.

Ford et al. [20] described an algorithm for the entailment based on induction of confirmation. We are interested in finding entailments between the constructs. These are not strictly logical or deductive entailments, but instead may be regarded as generalizations of the form "Every A is B", where A and B are constructs. Given some heuristic rules governing the permissible combinations of planes, the set of all possible generalizations derivable from the repertory grid data will be called the hypothetical set G. A generalization taken from G may then be evaluated by a rule of confirmation, in order to derive the degree of support provided by the evidence to generalization.

Based on the algorithm in [17] we generated the following rules from the grid shown in Fig.4:

IF defect=extremely circular **AND** defect=extremely greenish **THEN** defect= class_CI P= 1.0

The word extremely refers to the rating scaled as shown in Fig. 3.

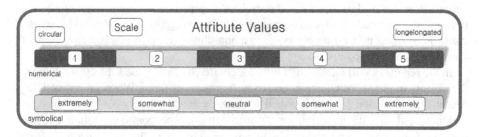

Fig. 3. Scale for Rating the Terms

The probability P gives a measure of degree of confirmation of the hypothesis.

The main advantage of the repertory grid technique is that it can be automated by a computer program. Other advantages are [16]:

1. It allows the expert to experiment with different approaches to his subject area without any inter-personal pressures. People are often reluctant to discuss the basis of their expertise, not because it is proprietary, but because they do not have an explicit model of it.
2. The on-line computer extraction of elements and constructs can be very much more effective than that by a person, because the computer system can analyzes the grid as it is being developed and feed back information to aid extraction.
3. The knowledge elicited by different subjects can be compared and evaluated. That helps to overcome the problem of "Who is the right expert?"

The method is a dichotomization method based on two poles. Sometimes it is not easy for a subject to decide what are the two poles. Figure 4 shows this. The construct "non_circular" and "circular" seem to be bipolar. The construct "random shape" and "longelongated shape" might not be bipolar. Maybe that is the reason that the importance of this construct is too low.

The result of the repertory grid methods is a hierarchy for the visual objects considered for classification and the constructs. The conceptual model of this hierarchy is hidden represented in the structure of the dendrogram. It needs to get manually labeled by an expert. As a result we would get a hierarchical conceptual model for the visual objects and the constructs.

Spatial relations between the objects can also be considered in the constructs. It allows us to not only have a feature-based representation, but also a feature and relation-based representation.

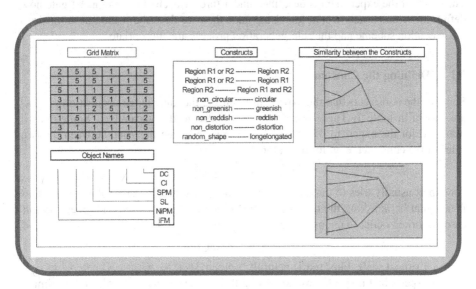

Fig. 4. Result from Repertory Grid Process

6   Experiment

The repertory grid technique was investigated under the aspect of whether the repertory grid is a good method for visual object classification or not. How can the tool be made more appropriate to visual inspection tasks and what kind of function is necessary. The experiment was carried out under the following conditions:

Five subjects involved with the same inspection task participated in the test as experts. Three of them were operators on the wafer-inspection line, one of them was an inspector and the fifth was a process engineer.

The products were highly structured objects made of different materials. Each product contained exactly one defect. Some defects were related to other part objects on the product and others were not related to each other. The defects varied in size, shape, surface curvature and surface imperfection.

A program was developed for the repertory grid technique and the entailment. The program runs on a PC and is based on C++. The experts were given an introduction to the repertory grid process. After the introduction the knowledge extraction was started. The experts had images of all defects and they were allowed to draw images on a paper block.

The acquisition process was done by think-aloud protocol analysis and observation. We considered 6 defects at a time.

7   Results

The result of the experiment is described under three aspects: First we investigate how definition of constructs works for visual taks, then we describe if the questionnaires, the execution scheme and the rule-generation process are appropriate.

7.1   Defining the Constructs

Some of the constructs that the experts used during the knowledge-acquisition process are shown in table 1. These learnt concepts make up the vocabulary. Sometimes it was hard for the expert to find the right features and even the scale between the two sides of the construct, e. g. for the color:

red-------------- green.

An expert usually tries to think about such a construct in terms of a color metric. This means that he is asking the question: what color is in between red and green? In this case it worked better to choose the construct as binary poles, like:

red---------------non_red,

or as a 3 valued scale, like "red(1)-neutral(2)-green(3)".

The experts had no problems expressing the intensity scale in terms of a minimum value white and a maximum value black. Here, the expert was able to think of the rating as the degree of membership of a defect at the right hand side or the left hand side of the construct, because most people are trained to express the intensity.

The experts required time to adapt to the questionnaire strategy and to the rating of elements to constructs. This indicates that an expert has no adhoc recallable structure for concepts like color and shape or other visual terms. So it seems to be better to give the expert an idea of what the constructs looks like in advance, or to provide a help function in the system. The following construct:

same shape---------------different shape

was chosen by an expert during a repertory grid process. The meaning of such a construct requires further explanation. During the extraction, the expert explained the meaning of the construct by drawing pictures and taking it as a reference to the construct. However, with the existing program, there was no way to capture his explanations.

The expert's thinking was stimulated by the method of choosing features to distinguish between elements in triads. Along with thinking about the objects and distinguishing features there was heavy an intense discussion going on, often accompanied by picture drawing. The discussions and pictures can be captured by the tool and stored into a data base as comments to the recent defect. They can be further evaluated by the experts when they have obtained new insights into the problem.

These examples show the need to capture explanations, discussions and pictures, because they contain important background material.

Table 1. Constructs

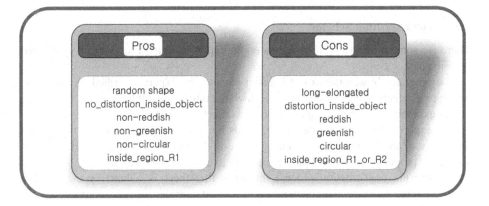

7.2   The Questionnaire

Buttler [16] has written about different interviewing techniques. He referred to the technique mentioned above with the following argument: "any of these interviewing techniques suffer from the problem that the number of pairs and triples of objects increase exponentially with the size of the set". Therefore, even if the interviewing methods can in principle discover the attribute structure of the set of objects, the process can be laborious and even impractical.

We discovered a problem with the above interview technique, which can make the process inefficient.

Consider the following example:
There are two categories of objects D1 D2 D3 D4 and D5 D6 d7 and these two categories are distinguished by one feature A1 and its opposite feature A2:

categories	C1	C2
defect types	D1 D2 D3 D4	D5 D6 D7
Feature	A1 'located in object region 1'	A2 'located in object-region 2'

Suppose the program first selects the triad D1, D2 and D5. Then the expert can distinguish elements D1 and D2 from D5 by selecting distinguishing constructs? A1 - A2. He is then asked to rate all elements (D1 trough D6) on the scale from A1 to A2. If the next randomly selected triad is D2, D6 and D, the expert could, in the absence of other distinguishing features, reuse the constructs A1 - A2 to distinguish D2 from D1 and D5.

The program does not recognize the fact that the same construct was reused. It requires the user to repeat the rating of all elements using A1 - A2. This is time-consuming and somewhat annoying for the user. We were able to observe that the repeated ratings were very consistent. The expert is not likely to change his mind, so by removing the repetition one will not lose any information.

Perhaps the program can be improved by: a) avoiding to ask for repeated ratings of the same construct and b) recognizing such major categories of elements and changing the triad selection strategy. In this case, selecting triads from within rather than from across, the major categories would improve in efficiency. Therefore we had to introduce a stopping rule into the program.

7.3   The Execution Scheme

There are some interesting observations on the repertory grid process that we want to describe here.

The program showed the expert 3 defect names. The expert described the property which two defects have in common and which the third does not have.

For example: "... shadow around defect..."

The expert confirms the meaning of words by showing the property at the image and/or by a? circle around the area of interest. If there is no image available that shows the property, he draws a picture in order to confirm the meaning.

If, after the confirmation process, he finds that there is a better word for the meaning of the property, then he would go back and consider the new word.

If he finds that the chosen property does not distinguish sufficiently the two defects from the third defect, he tries to find a better distinguishing property.

The experts never wanted to do the repertory grid process alone. They preferred to work in groups of two, because that gave them the possibility to discuss the defect image and to confirm their choices of constructs, features and rating.

7.4   The Rule Generation Process (Entailment)

It was assumed that a hierarchical representation of the knowledge would be best for the expert. But the expert preferred the rules generated by the entailment process. He could tell immediately whether he would accept the rules or not.

8   The Architecture of the Knowledge-Acquisition Tool

From the evaluation of the repertory grid method based on the inspection task described in the former section we developed the knowledge acquisition tool.

This tool provides the possibility of storing the verbal information as well as visual information.

This means that it has an interface to an image database from which the expert can easily and quickly retrieve and display the 3 defect images.

The verbal information about the defects is in generally the feature description. This minimal information can be kept in the repertory grid matrix.

The system also provides the expert with the ability to refine the features of the constructs after knowledge has been elicited. This allows one to generate more general descriptions. The system also provides a dictionary which helps the expert to find the right meaning of a word. New words can be entered in the dictionary with a description of the meaning of the word. That should give the expert the possibility to build up his own domain vocabulary. A basis for the content of the dictionary is given in section 3.

In addition, the system has an interactive graphic facility, as well as image-analysis algorithms and a property-scale library. The questionnaire of the repertory grid was adapted to the conversational strategy [21] of a CBR system, avoiding asking several times questions about the same objects.

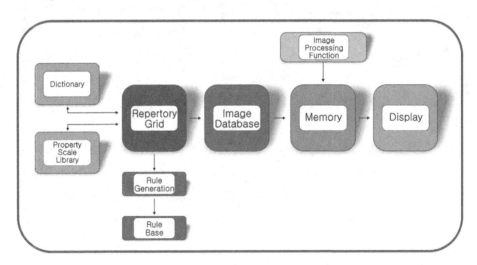

Fig. 5. Architecture

8.1   Interactive Graphic Facilities and Feature Extraction Algorithms

A person who is involved in image analysis will know the problem that arises when a new application has to be considered. The image material is new, meaning that the characteristic of the material is not well known. When image preprocessing and analysis algorithms are first applied without careful consideration of the image characteristics, surprising results can be discovered. There can be a region of a homogenous grey level or a high pass filter may show a line where no such region or line is expected. By carefully looking at the image, we will see that there is such actually corresponding information contained in the image, but we had no sense of it, because it was imbedded in the whole description until it became visible to us. The same seems to happen when two experts look at an image. Sometimes one of them sees more or different things.

8.2   The Property Library

A property-scale library is providing the user with a set of property scales for the rating of the constructs, which refers to certain concepts, for example shape, intensity and color.

Fig. 6 shows the concept of "intensity". As we can see, there exists no single scale for intensity. At the top of the hierarchy a decision between "constant" and "shaded" has to be made. The next hierarchy level under "constant" shows a three-pointed scale. Taking into account all hierarchy levels gives a 10 ten-pointed scale. The concept of "intensity" was learned based on the dual values for the grey levels.

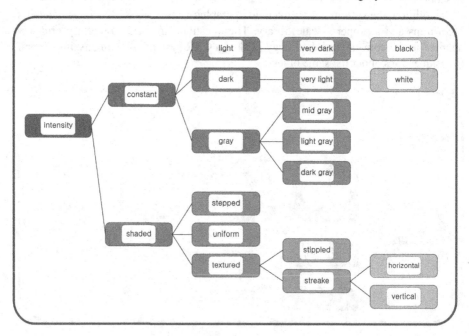

Fig. 6. Concept of "Intensity

It is clear that the system cannot provide all concepts initially. The system needs learning capabilities in order to learn the concepts with each newly discovered feature.

Our visual knowledge-acquisition tool *VKAT* is implemented on a PC in Borland C++.

9 Conclusion

Learning an ontology for us has to do with knowledge acquisition, in order to come up with the right vocabulary and the rules for the visual task. It also has to do with symbol-to-signal transformation which means developing the image-processing procedures for the extraction of the symbolic terms. We have shown that the repertory grid method combined with facilities, such as image processing procedures, image data base, dictionary and incremental learning and storage facilities, is a convenient semiautomatic tool for this task.

Knowledge modelling is important when the number of objects is large. The collected knowledge has to be available to the user in a convenient way. It should be possible to easily update the knowledge based on the already existing knowledge.

Our system *VKAT* handles that in an efficient way. All the elicited knowledge is available as constructs in the repertory grid matrix as well as in a dictionary. From the vocabulary the ontology can be constructed.

The above described technique can evaluate the elicited knowledge by cluster-analysis techniques. It helps to establish visual terms that have a high distinguishing ability. Thus the system developer gets a good idea of what he can expect for an output and what has to be done in order to make the system more efficient.

References

1. Reston, V.: BI-RADS® – Mammography, Breast Imaging Reporting and Data System Atlas (BI-RADS® Atlas), 4th edn. American College of Radiology (ACR) ©American College of Radiology (2003),
   http://www.acr.org/SecondaryMainMenuCategories/
   quality_safety/BIRADSAtlas/BIRADSAtlasexcerptedtext/
   BIRADSMammographyFourthEdition.aspx
2. Perner, P., Belikova, T.B., Yashunskaya, N.I.: Knowledge Acquisition by Decision Tree Induction for Interpretation of Digital Images in Radiology. In: Perner, P., Rosenfeld, A., Wang, P. (eds.) SSPR 1996. LNCS, vol. 1121, Springer, Heidelberg (1996)
3. Dom, B.E., Brecher, V.H., Bonner, R., Batchelder, J.S., Jaffe, R.S.: The P300: A system for automatic patterned wafer inspection. Machine Vision and Applications (4), 205–221 (1988)
4. Hedengren, K.: Methodology for Automatic Image-Based Inspection of Industrial Objects. In: Sanz, J.L.C. (ed.) Advances in Machine Vision, pp. 200–210. Springer, New York (1989)
5. Lundsteen, C., Gerdes, T., Philip, K.: Attributes for Pattern Recognition Selected By Stepwise Data Compression Supervised By Visual Classification. In: Kittler, J., Fu, K.S., Pau, L.F. (eds.) Pattern Recognition Theory and Applications, pp. 399–411 (1982)
6. Szafarska, E., et al.: A Image Archiv for Cataloging of Welding Seams. DGzfP-Jahrestagung 25.9.-27.9.89, Kiel, pp. 613–617 (Original is in German)
7. Perner, P.: Knowledge-Based Image Inspection System for Automatic Defect Recognition, Classification and Process Diagnosis. Machine Vision and Applications 7, 135–147 (1994)
8. Kelly, G.A.: The psychology of personal constructs, 2nd edn., vol. I, II. Norton, Routledge, New York, London (1955, 1991)

9. Manjunath, B.S., Salembier, P., Sikora, T.: Introduction to MPEG-7 Multimedia Content Description Interface. John Wiley & Sons Ltd. (2003)

10. Rao, A.R.: A Taxonomy for Texture Description and Identification. Springer, Berlin (1990)

11. Colantonio, S., Gurevich, I., Salvetti, O., Trusova, Y.: A Semantic Framework for Mining Novel Knowledge from Multimedia Data. In: Perner, P. (ed.) Machine Learning and Data Mining in Pattern Recognition, Poster Proceedings, pp. 62–67. IBaI (2009) ISBN 978-3-940501-04-2

12. Perner, P.: Image Mining: Issues, framework, a generic tool and its application to medical-image diagnosis. Journal Engineering Applications of Artificial Intelligence 15(2), 193–203

13. Clouard, R., Renouf, A., Revenu, M.: Human-computer interaction for the generation of image processing applications. Intern. Journal Human-Computer Studies 69, 201–219 (2011)

14. Wielinga, B., Sandberg, J., Schreiber, G.: Methods and Techniques for Knowledge Management: What has Knowledge Engineering to Offer? Expert Systems with Applications 13(1), 73–84

15. Finke, F.: Principles of Mental Imagery, pp. 89–90. MIT Press, Cambridge (1989)

16. Butler, K.A., Corter, J.E.: Use of Psychometric Tool for Knowledge Acquisition: A Case Study. In: Gale, W.A. (ed.) Artificial Intelligence and Statistics, pp. 293–319. Academic Press, Massachusetts (1986)

17. Boose, J.H., Shema, D.B., Bradshaw, J.M.: Recent progress in Aquinas: a knowledge acquisition workbench. Knowledge Acquisition 1, 185–214 (1989)

18. Anderberg, M.R.: Cluster Analysis for Applications, pp. 40–41. Academic Press, New York (1973)

19. Gaines, B.R., Shaw, M.L.G.: Induction of Inference Rules for Expert Systems. Fuzzy Systems 18, 315–328 (1986)

20. Ford, K.M., Petry, F.E., Adams-Webber, J.R., Chang, P.J.: An Approach to Knowledge Acquisition Based on the Structure of Personal Construct Systems. IEEE Trans. on Knowledge and Data Engineering 3(1), 78–88 (1991)

21. McSherry, D.: Increasing dialogue efficiency in case-based reasoning without loss of solution quality. In: Proceedings of the Eighteenth International Joint Conference on Artificial Intelligence, pp. 121–126. Morgan Kaufmann, San Francisco (2003)

Ontology and Algorithms Integration for Image Analysis

Sara Colantonio, Massimo Martinelli, and Ovidio Salvetti

Institute of Information Science and Technologies, ISTI-CNR
Via G. Moruzzi, 1 – 56124 Pisa, Italy
`name.surname@isti.cnr.it`

Abstract. Analyzing an image means extracting a number of relevant features for describing, meaningfully and concisely, image content. Usually, the selection of the features to be extracted depends on the image based task that has to be performed. This problem-dependency has caused the flourish of number and number of features in the literature, with a substantial disorganization of their introduction and definition. The idea behind the work reported in this paper is to make a step towards the systematization of the image feature domain by defining an ontological model, in which features and other concepts, relevant to feature definition and computation, are formally defined and catalogued. To this end, the *Image Feature Ontology* has been defined and is herein described. Such an ontology has the peculiarity of cataloguing features, modelling the image analysis domain and being integrated with a library of image processing algorithms, thus supplying functionalities for supporting feature selection and computation.

Keywords: Image Analysis, Feature Extraction, Ontologies.

1   Introduction

Image analysis for describing image visual content is the first step for any computerized application aimed at performing an image-based task, i.e., a task that relies on the interpretation and recognition of images.

Analyzing an image means extracting a number of meaningful and significant *features*, which express some relevant properties or characteristics of the image or portions of this. The significance of a feature is usually measured with respect to the task it is extracted for and a plethora of features have been introduced in the literature for solving specific problems. Actually, the features to be computed are commonly selected following an application-dependent approach; for instance:

- in image interpretation tasks, such as the diagnosis of leukemia in microscopy images of cell assays, features are computed to evaluate morpho-functional characteristics of the relevant image structures, i.e. the cells, in order to identify the malignant ones. The choice can fall on shape features (e.g., area, size, circularity,...) as well as textural features for the chromatin distribution inside the cell nucleus (e.g., *heterogeneity* and *margination* according to Mayall/Young's definition [1]), as has been done in [2];

E. Salerno, A.E. Çetin, and O. Salvetti (Eds.): MUSCLE 2011, LNCS 7252, pp. 17–29, 2012.

- in image recognition tasks, such as face detection, specific features are computed for detecting the peculiarities that identify a face, e.g., the Haar-like features as in [3];
- in content based image retrieval, features for summarizing and comparing image content are extracted: standardization approaches have been introduced to this end, i.e. MPEG-7 descriptors; though recently, low-bit rate local descriptors are experiencing wide diffusion, e.g., Compressed Histogram of Gradients descriptors (CHoG) or Scale Invariant Feature Transform (SIFT) [4].

The above examples highlight the heterogeneity of the types of features that can be extracted for describing image content: features are often defined and selected according to a problem-oriented strategy, i.e. *ad hoc* in light of the information considered relevant for the problem at hand. This peculiarity has caused the flourish of numbers of *ad hoc* features, many of which are not related to visually evident characteristics of images. Moreover, although several features can be computed in different ways, what often happens is that a precise definition of the features used is not even reported in the published work. Such lacks, of course, affect and prevent the verifiability and the reproducibility of the work done. A precise definition becomes especially necessary when complex features are used.

Some authors have tried to systematically review the literature on features [5], but a standardized vision is currently missing. A step towards the systematization of the image analysis domain can come from the definition of a formal model of features and their usage. The aim of this work is to present an ontological model, the *Image Feature Ontology (IFO)*, for structuring the domain of image features by an efficient definition of feature categories according to their generation, computation, and use. A great added value of this model is its integration with a library of algorithms for features computation, which assures the precise definition of each feature. The proposed approach currently contains plenty of features, but (of course) not all the features presented in the literature: the model is anyway easily upgradable and extensible since new features, operations and algorithms can easily be inserted in the corresponding categories. Actually, the feature domain organization supplied by the ontological model has a general flavour: the top level classes of the defined hierarchy model all the possible features (those already inserted as well as new ones): adding a new feature means just inserting a new "leaf" node to this hierarchy. The idea behind this work is to give a general structure to the feature domain, and populate it with an initial number of features, thus showing the feasibility of the approach. Extensions to model can easily follow, also by the image analysis community.

The model described is also a piece of an ontology-based framework which is being developed by the authors to support users in the development of image processing applications. The work reported in this paper is a fundamental component of this framework, which will be detailed in future works.

In the following sections, the problem of structuring image analysis domain is firstly introduced and similar works presented in the literature are reviewed (Sec. 2). Then, in section 3, the ontological model proposed is described, as well as its integration with the library of algorithms. Finally, a discussion about the model application is provided in section 4, with a sample approach. Conclusions and future works conclude the paper in section 5.

2   Problem Statement and Related Works

Systematizing the domain of image features means organizing the plethora of features that can be computed from an image by: (i) giving a precise definition of each feature; (ii) structuring the different features in different categories; (iii) emphasizing the peculiarities of features in accordance to how they are computed and used.
The final aim is to:

- standardize and homogenize the feature terminology;
- collect and disseminate structured classes of features;
- support the choice and computation of features according to a method-oriented strategy.

Though several definitions of feature have been presented in the literature [5], if the focus is put on the significance of features with respect to the image-based task to be performed, a feature can be defined as a characteristic or descriptor of image content, which can be suitably computed to represent or describe relevant peculiarities of images or portions of them. A first issue that can be derived from this definition is that a feature is specifically computed by means of dedicated operators and is, then, different from intrinsic properties of images, such as the colour model.

For organizing the features, several issues should be taken into account, among which: (i) features are computed differently according to the image type, e.g., to the colour model; (ii) they can be local or global; (iii) they can require the extraction of image relevant *structures* (i.e., image regions corresponding to visual objects contained in the image, which are relevant for the task at hand); (iv) they can describe different properties of images, e.g., intensity or colour characteristics, shape or texture properties of image regions; (v) they can require image pre-processing or the application of specific operators.

These issues can be mirrored in the definition of the features categories: in [5], for instance, features are categorized putting at the same level different, not related points of view, e.g., according to image characteristics (namely "information about the input image") as well as to image computation peculiarities (namely, "mathematical tools used for feature calculations"). This way, feature categorization is not organic and rather confusing.

In this work, the idea was to model all the above-reported issues by identifying the ones that correspond to a clear and well understandable categorization.

In this frame, ontologies appeared particularly suitable since they are a viable instrument for structuring the feature terminology, defining a feature taxonomy, and, most important, modelling their properties as well as all the relevant concepts involved in feature definition and computation. The latter is a fundamental issue since an ontology allows for the definition of a knowledge base that can be used for reasoning on the feature model produced.

The ontology-based approach has gained currency in recent years in the field of computer and information science. In particular, it has become popular in domains such as knowledge engineering, natural language processing, cooperative information systems, intelligent information integration, and knowledge management. An ontology can be defined as a tuple $\mathcal{O} = (C; R; \leq; \perp; \mathsf{I}; \sigma)$ where

- C is a finite set of concept symbols;
- R is a finite set of relation symbols;
- $\leq$ is a reflexive, transitive and anti-symmetric relation on C (*a partial order*);
- $\perp$ is a symmetric and irreflexive relation on C (*disjointness*);
- | is a symmetric relation on C (*coverage*);
- $\sigma: R \rightarrow C^+$ is the function assigning to each relation symbol its arity; the functor $(-)^+$ sends a set C to the set of finite tuples whose elements are in C [6].

This definition indicates the main elements of an ontology, namely the classes corresponding to the relevant concepts of a domain, the relations among these, and a class taxonomy specified by the partial order and with specific properties of disjointness and coverage. To use an ontology in a specific domain, the classes should be populated with instances or individuals. This induces the definition of a *populated ontology* as a tuple $\mathcal{P} = (C; R; \leq; \perp; |; \sigma)$ such that $C = (X; C; \models_C)$ and $R = (X^+; R; \models_R)$ are classifications and $\mathcal{O} = (C; R; \leq; \perp; |; \sigma)$ is an ontology. The ontology is correct when, for all $x, x_1, x_2, \ldots, x_n \in X$; $c; d \in C$; $r \in R$; and $\sigma(r) = \langle c_1; c_2; \ldots ; c_n \rangle$ the following holds true:

- if $x \models_C c$ and $c \leq d$, then $x \models_C d$;
- if $x \models_C c$ and $c \perp d$, then $x \not\models_C d$;
- if $c \mid d$, then $x \models_C c$ or $x \models_C d$;
- if $\langle x_1, \ldots, x_n \rangle \models_R r$ then $x_i \models_R c_i$, for all $i = 1, \ldots, n$.

So far, several ontologies have been defined for the image processing domain. Some of them contain a conceptualization of image features, but aim at specific image processing tasks. A *Visual Concept Ontology* has been proposed in [7] and recently re-used in [8] to define a general framework able to support the automatic generation of image processing applications (ranging from image pre-processing tasks, such as enhancement and restoration, to image segmentation tasks). Other ontologies have been developed to support semantic annotations of images [9, 10]. These are mainly based on the translations of the MPEG-7 descriptors into an ontological model, with few extensions [11]. All these works have different drawbacks, since they are either too general, since try to model all the image processing tasks, or too focused, since limited to a set of features useful especially in image retrieval. Moreover, it is difficult to find in the literature an *operative* approach that combines a descriptive ontology with computation algorithms. Recently, a kind of integration has been proposed [12], but, again, this is focused on MPEG-7 descriptors.

The work reported in this paper aims at defining the IFO model specifically dedicated to the systematization of image analysis domain, and at integrating this with a library of algorithms. Even though the ontology contains a preliminary set of features, the proposed approach shows the feasibility of the idea that drives the domain organization. All the ontologies presented in the literature have been accurately reviewed to analyze the feature classification they propose and the feature they consider, so as to take into account the significant pieces of information. None of them was suitable to be reused as is, for the purpose of this work. In the following sections, an insight into the developed IFO is reported.

3   The Image Feature Ontology

Several important issues were taken into account and different choices were made to develop the IFO model. These can be ascribed to the different facets of the image analysis elements, namely, the image characteristics, the feature properties, and the feature computation requirements.

More precisely, the following conditions were considered as peculiarities of the feature domain and mapped on the model requirements and/or desiderata; some of them derive from the definition of feature as given in the previous section:

- Image level: (*i*) Intrinsic image characteristics, such as the colour model, are not to be considered as features; (*ii*) different image types should be considered as well as the correspondence between these image types and the features selected for them;
- Feature level: (*i*) Features can be local or global, i.e. computed on the entire image or portions of this; (*ii*) Some features can be computed in different ways, e.g., the area of a region can be computed as the number of pixels or as the area of an irregular polygon (i.e., as the sum of several triangles that compose the polygon); (*iii*) Different image characteristics are usually described by different features, but in some cases the same feature can be used to describe different properties, e.g., the moments of the intensity or colour distribution (i.e., the histogram) are commonly used to describe intensity/colour properties as well as textural properties; (*iv*) Features should be defined precisely, avoiding ambiguity between elementary image components (i.e., edge, points of interest, regions of interest,...) and features computed on them; (*v*) Important characteristics of features, such as their invariance to possible image transformations, should be pointed out and highlighted to ease the selection of features for different tasks;
- Feature computation level: the relevant concepts pertaining feature computation should be modelled to both support feature selection and assure the computation of the selected features. This translates into the following conditions: (*i*) Operational constraints and requirements should be explicitly coded; (*ii*) Links to the algorithms for computing the different features should be modelled within the ontological model.

Starting from these conditions, IFO was defined first of all by analyzing the literature sources about features and their classification. Without the claim of including all the features, well known monographs and image analysis reviews [1, 5] were analyzed as well as a number of papers presented in the literature.

Different possible feature categorizations resulted from this analysis, as well as a remarkable complexity of feature definition. Several modelling choices appeared compulsory. More precisely, the following decisions were made:

- At a high level of the feature taxonomy, features were classified according to a well recognized image property they describe; this is particularly suitable to allow a better comprehension of the categorization by experts and non experts, and makes it easier to support selection of features according to the image properties relevant to the problem at hand;

- At the inner levels of the taxonomy, features were classified according to the pre-computation they require, e.g., image primitive elements such as edges, contours, or image basic functions, such as the histogram and the co-occurrence matrix;
- A distinction was made among features and visual primitives, such edges, region of interests, skeletons, and so on so forth;
- All the computation requirements, necessary to evaluate specific features, was highlighted and specific concepts was introduced.

3.1 Ontology Components

According to the above choices, an ontological model was defined with the general sketch reported in Fig. 1, which shows the main concepts and the relations among them.

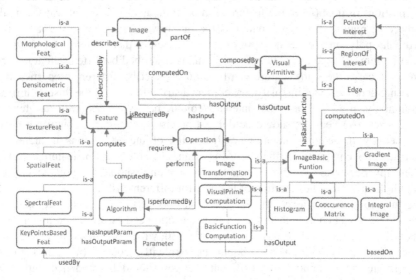

Fig. 1. A high level vision of the *Image Feature Ontology*

For better describing the model, a detailed presentation of the main concepts, taxonomies, properties and relations is given in the following.

Image Related Concepts

Image is obviously a key concept of the model and has a number of properties: hasColumns and hasRows, which indicate image dimensions; hasColourModel, which reports image colour model (with value in the set of RGB, HSV, HSI, CMYK, BW GreyLevel); hasOrigin, which indicates how the image has been acquired or obtained; hasBitResolution, which reports the image quantization rate; hasAnnotation which reports possible semantic information associated to the image. Among these properties, a particular focus is given to ColourModel, since it is used for a first distinction among the features that can be computed.

An Image can be composedBy some VisualPrimitives, which can be (i) one or more PointOfInterest, characterized by hasXPosition and hasYPosition; (ii) one or more RegionOfInterest and (iii) one or more Edge. These concepts were modelled according to a point-based representation. In particular, a sequence of points was introduced by defining the concept SequentialPoint: this locates a point into the sequence, and is linked to the following point by the hasNext relation. An Edge is linked to its sequence of points by the consistingOf relation. For RegionOfInterest, Contour and Skeleton representations were introduced, both consisting of sequences of points (i.e., consistOf and formedBy relations). The concept BoundingBox was also defined and related to RegionOfInterest (i.e. hasBoundingBox relation); this concept has four relations with the concept PointOfInterest to consider just the four vertexes of the box (i.e., hasHighLeftVertex, hasHighRightVertex, hasLowLeftVertex, hasLowRightVertex). Finally, a RegionOfInterest hasCentroid in one PointOfInterest. Fig. 2 summarizes these concepts and relations.

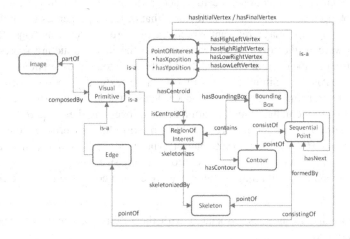

Fig. 2. Concepts and relations that pertain the VisualPrimitive concept

An Image is also describedBy one or more Feature, it is the input argument of processing functions of the class Operation and the output of ImageTransformation operations.

Features Related Concepts

The concept Feature was introduced to model the feature as introduced in the previous sections, hence a Feature describes an Image and has several proprieties: i.e., hasFormalDefinition to report a precise, mathematical description of the feature; hasInformalDefinition, to describe, when possible, the visual interpretation of the feature; hasApplicabilityLevel, to indicate if the feature is local, global or both; hasReference, to indicate the bibliographic source of the feature; hasBeenUsedFor, to report a number of image based tasks the feature has been used for in the literature; correspondsTo, to report the equivalent descriptor of MPEG-7, if any; alsoKnownAs, to report different names for the same feature. Currently the tasks are not explicitly formalized into the ontological model, but in the next versions a specific concept Task will be modelled, to support the automated selection of features.

A classification of features was identified and represented in the model taxonomy. The strategy adopted to define such a classification followed both a top down and a bottom up approach. An initial, large enough, set of features was identified. These were, from one side, grouped according to classifications founde in the literature, and, from the other side, they were analyzed to identify a possible classification. These two different classifications were, then, compared to evaluate their coverage and plausibilty, and a compromise between the different solutions was chosen.

The selected categorization divides the features according to the image properties they describe; in particular, the following high level categories were identified:

- DensitometricFeature, composed by features that describe colour or intensity properties. These were further divided according to the image colour model, so as to simplify the selection of features according to image typologies;
- MorphologicalFeature, consisting in features that describe properties about the shape and topology of image relevant structures that are extracted as RegionOfInterest. These were further divided in accordance to the image structure representation they use, i.e., the contour representation or the skeleton representation;
- TextureFeature, containing features that describe textural properties, of the entire image or portion of it, i.e., a RegionOfInterest. These were further categorized according to the image transformation or pre-processing they require;
- SpatialFeature, collecting features that state relationship properties among relevant structures extracted as RegionOfInterest. These derived from [7] and take into account topological and extrinsic relations among structures;
- KeyPointsFeature, containing features that are computed to describe specific points of the image.

Fig. 3 shows the taxonomies associated with the different categories. Some issues should be pointed out:

- the different features were reported into the model as concepts; the individuals belonging to the corresponding classes are instances of the specific features computed on a specific image, which is, in turn, an instance of the class Image;
- the above classification produces classes that are not disjoint, hence there can be several features that belong to more than one sub-taxonomies, e.g., all the features computed on the image histogram can be classified as DensitometricFeature as well as TextureFeature. The choice was to insert different concepts for the feature in the different taxonomies and then explicitly state the equivalence among the corresponding concepts, e.g., IntensityMean isEquivalentTo TexturalMean;
- the concepts belonging to the different sub-taxonomies were further specialized by introducing more properties, for instance the RegionBasedFeature has the property invariantTo, which specifies the image transformations the feature is invariant to.

The Feature to be computed can require an Operation, which can be an ImageTransformation (e.g., a ColourMapping, FourierTrasform), or VisualPrimitiveComputation for the extraction of a VisualPrimitive (e.g., SobelEdgeExtraction to extract a series of Edge), or BasicFunctionComputation for calculating a BasicFunction (e.g., GreyLevelHistogramComputation to compute the GreyLevelHistogram). In particular, among the BasicFunction, a distinction was made according to the image colour models, hence the concept Histogram, for instance, was further specialized in GreyLevelHistogram and ColourHistogram.

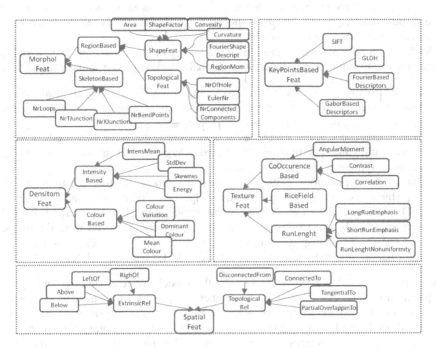

Fig. 3. The sub-taxonomies of the concepts: DensitometricFeature, MorphologicalFeature, TextureFeature, SpatialFeature and KeyPointBasedFeature (GLOH stays for gradient location and orientation histogram). The arches depicted correspond all to the is-a relation.

Finally, the concept Algorithm was introduced to model the computation of any Feature (as stated by the relation computes) and any Operation (as stated by the relation performs). This is a key concept of the ontology and was characterized with the properties hasClass and hasMethod, which correspond, respectively, to the class and method of the library of algorithms. A concept Parameter was introduced to model the algorithms' parameters, characterized by the property hasType. Two relations, i.e., hasInputParameter and hasOutputParameter, connect Algorithm to this concept, as in [13]. A Feature as well as an Operation can be computed by one or more algorithms. Currently, Algorithm is not further specified and the individuals of this class correspond to the algorithms of an image processing library adopted to make the ontological model operative.

4   Ontological Model Development and Application

The *Image Feature Ontology* was developed by using the *de facto* standard language for ontology definition, i.e. the Ontology Web Language (OWL), in particular OWL2-DL[1].

Axioms and sub-properties, peculiarities of this language, were exploited to specialize *object-properties*, e.g., IntensityMean for an image that hasColorModel GreyLevel requires the computation of GreyLevelHistogram. Currently, the ontology consists of 353 concepts, and 168 properties. Among concepts, 180 are features and 86 are operations required by their computation. To implement these, 266 individuals of the class Algorithm were introduced.

Such an ontology and its integration with a library of algorithms represent a viable apparatus able to supply several, useful functionalities. IFO can be actually used by

- image processing experts, when developing image processing applications, to browse the "features catalogue" and find some features relevant to the application at hand, as well as their implementation;
- non experts users, when searching for interesting features for their needs: feature classification and specification can be used to suggest relevant features according to image typology, for instance;
- autonomic applications, within a dedicated framework, for the discovery of applicable features according to image properties. The most advanced usage of IFO would be the development of an application able to interpret natural language user's requests about the information to be extracted from the image and supply, in correspondence, the list of extractable features to obtain such information, plus the operations to be performed to compute such features.

Currently, a system has been defined to test the model (by command line, without a graphical user interface), based on a number of queries (implemented in *Simple Protocol and RDF Query Language*[2] – SPARQL) and on the integration of the library of algorithms. ImageJ[3] was selected as a Java-based, extensible library, with a number of image processing functions already developed. The integration of the library was not a trivial task, since it required the online automated formulation of a call to one algorithm, by just knowing its ontology specification, i.e., algorithm name and method, list of input and output parameters. The *Reflection* programming technique has been exploited to this end.

A simple scenario can be reported to illustrate the system currently developed for testing IFO. A series of steps can be identified as shown in Fig. 4.

- Initially, at the step 1, the user is asked to select an image. Once an image has been selected, the system automatically creates a corresponding individual of the class Image into IFO.

---

[1] http://www.w3.org/TR/owl2-overview/

[2] http://www.w3.org/TR/rdf-sparql-query/

[3] http://rsbweb.nih.gov/ij/

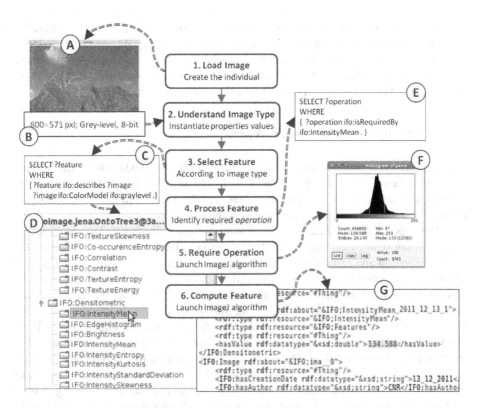

Fig. 4. The sequence of steps for the sample scenario. (A) the selected image; (B) image characteristics used to instantiate the image individual's properties; (C) the query launched to select the features computable according to image colour model; (D) an excerpt of the list of features resulting from the query; among them IntensityMean is selected; (E) the query launched to check the operation required by IntensityMean; (F) the results of the histogram computation by ImageJ; (G) the feature individual instantiated with the computed value.

- Then, at step 2, by a first call to ImageJ, the system obtains the basic characteristics of the image (i.e., size, resolution, colour model) and instantiates correspondingly the individual's properties.

- At step 3, according to the image properties, the list of computable features is computed by applying a selection query: for instance, according to the image colour model, a query is applied to select all the features that can be computed on a *GreyLevel* image (see Fig.4).

- At step 4, when one of these features is selected, the system discovers the operations required by the feature (see Fig.4 (C) for the query). If one operation has not been performed yet (i.e., the corresponding individual does not exist into the ontology), the system requires its execution to the user. For instance, supposing to select IntensityMean, the system discovers that the GreyLevelHistogram should be computed.

- Then, after approval by the user, in step 5, the system calls ImageJ to compute it. This supplies the histogram, displayed as shown in Fig. 4, and computes the mean. The resulting value is used to create the IFO individual IntensityMean related to the image selected (in Fig.4, this individual named IntensityMean_2011_12_13_1 is shown with its value instantiated with the number obtained).

5   Conclusions and Future Works

Extracting relevant features to describe image content is a fundamental step of any application based on the interpretation of such content. Usually, the selection of the features to be extracted is problem-dependent, and this resulted in the proliferation of numbers and numbers of features presented in the literature.

In this paper, a systematization of the feature domain is presented. The main idea is to develop an ontological model, the Image Feature Ontology (IFO), to standardize feature definition and computation. A classification of features according to the image properties they describe was conceived, and a number of features were categorized. Besides a feature classification, IFO also models the relevant concepts necessary to compute the features. A fundamental issue, in this perspective, is the integration of the ontological model with a library of algorithms for feature computation. This grants the effectiveness and functionality of the ontology. The work is ongoing, since the feature set is not complete. Moreover, for a validation of the proposed model, the IFO will be sent to experts in image processing, in order to get suggestions about its structure as well as new features and operations. The authors are also planning to use IFO model in a semantic framework able to support the development of image based applications (see [14]).

References

1. Rodenacker, K., Bengtsson, E.: A feature set for cytometry on digitized microscopic images. Analytical Cellular Pathology 25, 1–36 (2003)
2. Colantonio, S., Perner, P., Salvetti, O.: Diagnosis of lymphatic tumours by case-based reasoning on microscopic images. Trans.on Case-Based Reas. 2(1), 29–40 (2009)
3. Viola, P., Jones, M.: Rapid object detection using boosted cascade of simple features. In: IEEE CVPR, pp. 511–518. IEEE Press, New York (2001)
4. Chandrasekhar, V., et al.: Comparison of local feature descriptors for mobile visual search. In: ICIP, pp. 3885–3888 (2010)
5. Gurevich, I.B., Koryabkina, I.V.: Comparative Analysis and Classification of Features for Image Models. Pattern Rec. and Image Analysis 16(3), 265–297 (2006)
6. Kalfoglou, Y., Schorlemmer, M.: IF-Map: An Ontology-Mapping Method Based on Information-Flow Theory. In: Spaccapietra, S., March, S., Aberer, K. (eds.) Journal on Data Semantics I. LNCS, vol. 2800, pp. 98–127. Springer, Heidelberg (2003)
7. Maillot, N., Thonnat, M.: Ontology-based complex object recognition. Image and Vision Comput. 26(1), 102–113 (2008)
8. Clouard, R., Renouf, A., Revenu, M.: Human-computer interaction for the generation of image processing applications. Int. J. of Human-Computer Studies 69(4), 201–219 (2011)

9. Hunter, J.: Adding Multimedia to the Semantic Web - Building an MPEG-7 Ontology. In: Proc. Int. Semantic Web Working Symposium (SWWS) Stanford, July 30-August 1 (2001)
10. Arndt, R., Troncy, R., Staab, S., Hardman, L., Vacura, M.: COMM: Designing a Well-Founded Multimedia Ontology for the Web. In: Aberer, K., Choi, K.-S., Noy, N., Allemang, D., Lee, K.-I., Nixon, L.J.B., Golbeck, J., Mika, P., Maynard, D., Mizoguchi, R., Schreiber, G., Cudré-Mauroux, P. (eds.) ASWC 2007 and ISWC 2007. LNCS, vol. 4825, pp. 30–43. Springer, Heidelberg (2007)
11. Hollink, L., Little, S., Hunter, J.: Evaluating the Application of Semantic Inferencing Rules to Image Annotation. In: 3rd Int. Conf. on Know. Capt., KCAP 2005, Banff (October 2005)
12. Vacura, M., Svatek, V., Saathoff, C., Franz, T., Troncy, R.: Describing low-level image features using the COMM ontology. In: 15th IEEE Conf. on ICIP, pp. 49–52. IEEE Press (2008)
13. Asirelli, P., Little, S., Martinelli, M., Salvetti, O.: - Algorithm representation Use Case. In: Multimedia Annot. Interop. Framework. Multimedia Semantics Incubator Group (2006)
14. Colantonio, S., Salvetti, O., Gurevich, I.B., Trusova, Y.: An ontological framework for media analysis and mining. Pattern Rec. and Im. An. 19(2), 221–230 (2009)

EmotiWord: Affective Lexicon Creation with Application to Interaction and Multimedia Data

Nikos Malandrakis[1], Alexandros Potamianos[1], Elias Iosif[1],
and Shrikanth Narayanan[2]

[1] Dept. of ECE, Technical Univ. of Crete, 73100 Chania, Greece
{nmalandrakis,potam,iosife}@telecom.tuc.gr
[2] SAIL Lab, Dept. of EE, Univ. of Southern California, Los Angeles, CA 90089, USA
shri@sipi.usc.edu

Abstract. We present a fully automated algorithm for expanding an affective lexicon with new entries. Continuous valence ratings are estimated for unseen words using the underlying assumption that semantic similarity implies affective similarity. Starting from a set of manually annotated words, a linear affective model is trained using the least mean squares algorithm followed by feature selection. The proposed algorithm performs very well on reproducing the valence ratings of the Affective Norms for English Words (ANEW) and General Inquirer datasets. We then propose three simple linear and non-linear fusion schemes for investigating how lexical valence scores can be combined to produce sentence-level scores. These methods are tested on a sentence rating task of the SemEval 2007 corpus, on the ChIMP politeness and frustration detection dialogue task and on a movie subtitle polarity detection task.

Keywords: language understanding, emotion, affect, affective lexicon.

1 Introduction

Affective text analysis, the analysis of the emotional content of lexical information is an open research problem that is very relevant for numerous natural language processing, web and multimodal dialogue applications. Emotion recognition from multimedia streams (audio, video, text) and emotion recognition of users of interactive applications is another area where the affective analysis of text plays an important, yet still limited role [10,9,2].

The requirements of different applications lead to the definition of affective sub-tasks, such as emotional category labeling (assigning label(s) to text fragments, e.g., "sad"), polarity recognition (classifying into positive or negative) and subjectivity identification (separating subjective from objective statements). The wide-range of application scenarios and affective tasks has lead to the fragmentation of research effort. The first step towards a general task-independent solution to affective text analysis is the creation of an appropriate affective lexicon, i.e., a resource mapping each word (or term) to a set of affective ratings.

E. Salerno, A.E. Çetin, and O. Salvetti (Eds.): MUSCLE 2011, LNCS 7252, pp. 30–41, 2012.

A number of affective lexicons for English have been manually created, such as the General Inquirer [15], and Affective norms for English Words (ANEW) [3]. These lexica, however, fail to provide the required vocabulary coverage; the negative and positive classes of the General Inquirer contain just 3600 words, while ANEW provides ratings for just 1034 words. Therefore, computational methods are necessary to create or expand an already existing lexicon. Well-known lexica resulting from such methods are SentiWordNet [5] and WORDNET AFFECT [17]. However, such efforts still suffer from limited coverage.

A variety of methods have been proposed for expanding known lexica with continuous affective scores, or, simply for assigning binary "positive - negative" labels to new words, also known as semantic orientation [7]. The underlying assumption at the core of these methods is that *semantic similarity can be translated to affective similarity*. Therefore given some metric of the similarity between two words one may derive the similarity between their affective ratings. A recent survey of metrics of semantic similarity is presented in [8]. The semantic similarity approach, pioneered in [20], uses a set of words with known affective ratings, usually referred as *seed words*, as a starting point. Then the semantic similarity between each new word in the lexicon and the seed words is computed and used to estimate the affective rating of new words. Various methods have been proposed to select the initial set of words: seed words may be the lexical labels of affective categories ("anger","happiness"), small sets of words with unambiguous (affective) meaning or, even, all words in an affective lexicon. Having a set of seed words and the appropriate similarity measure, the next step is devising a method of combining them to estimate the affective score or category.

Once the affective lexicon has been expanded to include all words in our vocabulary, the next step is the combination of word ratings to estimate affective scores for larger lexical units, phrases or sentences. Initially the subset of affect-bearing words has to be selected, depending on their part-of-speech tags [4], affective rating and/or sentence structure [1]. Then word-level ratings are combined, typically in a simple fashion, such as the arithmetic mean. More complex approaches take into account sentence structure, word/phrase level interactions such as valence shifters [14] and large sets of manually created rules [4,1].

In this paper, we aim to create an affective lexicon with fine-grained/pseudo-continuous valence ratings. This lexicon can be readily expanded to cover unseen words with no need to consult any linguistic resources. The work builds on [20]. The proposed method requires only a small number (a few hundred) labeled seed words and a web search engine to estimate the similarity between the seed and unseen words. Further, to improve the quality of the affective lexicon we propose a machine learning approach for training a valence estimator. The affective lexicon created is evaluated against manually labeled corpora both at the word and the sentence level, achieving state-of-the-art results despite the lack of underlying syntactic or pragmatic information in our model. We also investigate the use of small amounts of in-domain data for adapting the affective models. Results show that domain-independent models perform very well for certain tasks, e.g., for frustration detection in spoken dialogue systems.

2   Affective Rating Computation

Similarly to [20], we start from an existing, manually annotated lexicon. Then we automatically select a subset of seed words, using established feature selection algorithms. The rating (in our case valence) for an unseen word is estimated as the linear combination of the ratings of seed words weighted by the semantic similarity between the unseen and seed words. Semantic similarity weights are computed using web hit-based metrics. In addition, a linear weight is used that regulates the contribution of each seed word in the valence computation formula. The weight of each seed word is optimized to minimize the mean square estimation error over all words in the training set.

Introducing a trainable weight for each seed word is motivated by the fact that semantic similarity does not fully capture the relevance of a seed word for valence computation. For instance, consider an unseen word and a lexicon that consists of two seed words that are equally (semantically) similar to the unseen word. Based on the assumption that semantic similarity implies affective similarity both seed words should be assigned the same feature weight. However, there is a wide range of factors affecting the relevance of each seed word, e.g., words that have high affective variance (many affective senses) might prove to be worse features that affectively unambiguous words. Other factors might include the mean valence of seed words and the degree of centrality (whether they are indicative samples of their affective area). Instead of evaluating the effect of each factor separately, we choose to use machine learning to estimate a single weight per seed word using Least Mean Squares estimation (LMS).

2.1   Word Level Tagging - Metrics of Semantic Similarity

We aim at characterizing the affective content of words in a continuous valence range of $[-1, 1]$ (from very negative to very positive), *from the reader perspective*. We hypothesize that the valence of a word can be estimated as a linear combination of its semantic similarities to a set of seed words and the valence ratings of these words, as follows:

$$\hat{v}(w_j) = a_0 + \sum_{n=1}^{N} a_i \, v(w_i) \, d(w_i, w_j), \tag{1}$$

where w_j is the word we mean to characterize, $w_1...w_N$ are the seed words, $v(w_i)$ is the valence rating for seed word w_i, a_i is the weight corresponding to word w_i (that is estimated as described next), and $d(w_i, w_j)$ is a measure of semantic similarity between words w_i and w_j.

Assume a training corpus of K words with known ratings and a set of $N < K$ seed words for which we need to estimate weights a_i, we can use (1) to create a system of K linear equations with $N + 1$ unknown variables as shown next:

$$
\begin{bmatrix}
1 & d(w_1, w_1)v(w_1) & \cdots & d(w_1, w_N)v(w_N) \\
\vdots & \vdots & \vdots & \vdots \\
1 & d(w_K, w_1)v(w_1) & \cdots & d(w_K, w_N)v(w_N)
\end{bmatrix}
\cdot
\begin{bmatrix}
a_0 \\
a_1 \\
\vdots \\
a_N
\end{bmatrix}
=
\begin{bmatrix}
1 \\
v(w_1) \\
\vdots \\
v(w_K)
\end{bmatrix}
\tag{2}
$$

where $a_1...a_N$ are the N seed word weights and a_0 is an additional parameter that corrects for affective bias in the seed word set. The optimal values of these parameters can be estimated using LMS. Once the weights of the seed words are estimated the valence of an unseen word w_j can be computed using (1).

To select N seeds out of K labeled words, we need to perform feature selection or simply randomly select a subset. Here, we rank all K words by their "worth" as features, using a wrapper feature selector working on a "best-first" forward selection strategy. The performance metric used for feature evaluation is Mean Square Error. Thus, we simply pick the N seeds that are best at predicting the affective rating of the rest of the training data. The seeds typically correspond to 10% to 50% of the training set (for more details see also Sections 4 and 5).

The valence estimator defined in (1) employs a metric $d(w_i, w_j)$ that computes the semantic similarity between words w_i and w_j. In this work, we use hit-based similarity metrics that estimate the similarity between two words/terms using the frequency of co-occurence within larger lexical units (sentences, documents). The underlying assumption is that terms that often co-occur in documents are likely to be related. A popular method to estimate co-occurrence is to pose conjunctive queries including both terms to a web search engine; the number of returned hits is an estimate of the frequency of co-occurrence [8]. Hit-based metrics do not depend on any language resources, e.g., ontologies, and do not require downloading documents or snippets, as is the case for context-based semantic similarities. In this work, we employ four well-established hit-based metrics namely, Dice coefficient, Jaccard coefficient, point-wise mutual information (PMI), and Google-based Semantic Relatedness (Google). Due to space limitations details are omitted, but the reader may consult [8].

2.2 Sentence Level Tagging

The principle of compositionality [13] states that the meaning of a phrase or sentence is the sum of the meaning of its parts. The generalization of the principle of compositionality to affect could be interpreted as follows: to compute the valence of a sentence simply take the average valence of the words in that sentence. The affective content of a sentence $s = w_1 w_2 ... w_N$ in the simple linear model is:

$$
v_1(s) = \frac{1}{N} \sum_{i=1}^{N} v(w_i). \tag{3}
$$

This simple linear fusion may prove to be inadequate for affective interpretation given that non-linear affective interaction between words (especially adjacent words) in the same sentence is common. It also weights equally words that have a strong and weak affective content and tends to give lower absolute valence scores to sentences that contain many neutral (non-content) words. Thus we also consider a normalized weighted average, in which words that have high absolute valence values are weighted more, as follows:

$$v_2(s) = \frac{1}{\sum\limits_{i=1}^{N} |v(w_i)|} \sum_{i=1}^{N} v(w_i)^2 \cdot \text{sign}(v(w_i)), \qquad (4)$$

where sign(.) is the signum function. Alternatively we consider non-linear fusion, in which the word with the highest absolute valence value dominates the meaning of the sentence, as follows. Equations (3), (4) and (5) are the fusion schemes used in the following experiments, where they are referred to as: average (avg), weighted average (w.avg) and maximum (max).

$$v_3(s) = \max_i(|v(w_i)|) \cdot \text{sign}(v(w_z)), \quad \text{where} \quad z = \arg\max_i(|v(w_i)|) \qquad (5)$$

3 Corpora and Experimental Procedure

Experimental Corpora: (1) ANEW: The main corpus used for creating the affective lexicon is the ANEW dataset. It consists of 1034 words, rated in 3 continuous dimensions of arousal, valence and dominance. In this work, we only use the valence ratings provided. ANEW was used for both training (seed words) and evaluation using cross-validation (as outlined below). (2) General Inquirer: The second corpus used for evaluation of the affective lexicon creation algorithm is the General Inquirer corpus that contains 2005 negative and 1636 positive words. It was created by merging words with multiple entries in the original lists of 2293 negative and 1914 positive words. It is comparable to the dataset used in [20,21]. (3) SemEval: For sentence level tagging evaluation (no training is performed here, only testing) the SemEval 2007: Task 14 corpus is used [16]. It contains 1000 news headlines manually rated in a fine-grained valence scale $[-100, 100]$, which is rescaled to a $[-1, 1]$ for our experiments. (4) Subtitles: For the movie subtitle evaluation task, we use the subtitles of the corpus presented in [11]. It contains the subtitles of twelve thirty minute movie clips, a total of 5388 sentences. Start and end times of each utterance were extracted from the subtitles and each sentence was given a continuous valence rating equal to the average of the multimodal affective curve for the duration of the utterance. (5) ChIMP: The ChIMP database was used to evaluate the method on spontaneous spoken dialog interaction. It contains 15585 manually annotated spoken utterances, with each utterance labeled with one of three emotional state tags: neutral, polite, and frustrated [22].

Semantic Similarity Computation: In our experiments we utilized four different similarity metrics based on web co-occurrence, namely, Dice coefficient,

Jaccard coefficient, point-wise mutual information (PMI) and Google-based Semantic Relatedness (Google). To compute the co-occurrence hit-count we use conjunctive "AND" web queries [8] so we are looking for co-occurrence anywhere inside a web document. All queries were performed using the Yahoo! search engine. The number of seed words determines the number of queries that will be required. Assuming that N seed words are selected, $N + 1$ queries will be required to rate each new word.

3.1 Affective Lexicon Creation

The ANEW dataset was used for both training and testing using 5-fold cross-validation. On each fold, 80% of the ANEW words was used for training and 20% for evaluation. For each fold, the words in the training set were ranked based on their value as features using a wrapper (for feature selection) on a 3-fold cross-validation prediction experiment conducted within the training set. On each fold 66% of the training data were used to predict the ratings of the remaining 33%, using our method. The selection started with no seed words, then more seed words were added iteratively - in each iteration all non-seeded words in the training set were tested as seeds, then the one that produced the lowest mean square error was added to the previous seed word set. The order in which words were added to the selection seed word set was their ranking. Then the N first were used as seed words. We provide results for a wide range of N values, from 1 to 500 features. Then the semantic similarity between each of the N features and each of the words in the test set ("unseen" words) was computed, as discussed in the previous section. Next for each value of N, the optimal weights of the linear equation system matrix in (2) were estimated using LMS. For each word in the test set the valence ratings were computed using (1).

A toy training example using $N = 10$ features and the Google Semantic Relatedness hit-based metric is shown in Table 3.1. The seed words are presented in the order they were selected by the wrapper feature selection method. The last row in the table corresponds to the bias term a_0 in (1) that takes a small positive value. The third column $v(w_i)$ shows the manually annotated valence of word w_i, while the fourth column a_i shows the corresponding linear weight computed by the LMS algorithm. Their product (final column) $v(w_i) \times a_i$ is a measure of the affective "shift" of the valence of each word per "unit of similarity" to that seed word (see also (1)).

Table 1. Training sample using 10 seed words

Order	w_i	$v(w_i)$	a_i	$v(w_i) \times a_i$	Order	w_i	$v(w_i)$	a_i	$v(w_i) \times a_i$
1	mutilate	-0.8	0.75	-0.60	6	misery	-0.77	8.05	-6.20
2	intimate	0.65	3.74	2.43	7	joyful	0.81	6.4	5.18
3	poison	-0.76	5.15	-3.91	8	optimism	0.49	7.14	3.50
4	bankrupt	-0.75	5.94	-4.46	9	loneliness	-0.85	3.08	-2.62
5	passion	0.76	4.77	3.63	10	orgasm	0.83	2.16	1.79
					-	w_0 (offset)	1	0.28	0.28

In addition to the ANEW corpus, the General Inquirer corpus was used (only) for testing. The features and corresponding linear weights were trained on the (whole of the) ANEW corpus and used as seeds to estimate continuous valence ratings for the General Inquirer corpus. Our goal here was not only to evaluate the proposed algorithm, but also investigate whether using seeds from one annotated corpus can robustly estimate valence ratings in another corpus.

The following objective evaluation metrics were used to measure the performance of the affective lexicon expansion algorithm: (i) Pearson correlation between the manually labeled and automatically computed valence ratings, (ii) mean square error (assuming that the manually labeled scores were the ground truth) and (iii) binary classification accuracy of positive vs negative relations, i.e., continuous ratings are produced, converted to binary decisions and compared to the ground truth.

3.2 Sentence Level Tagging

The SemEval 2007: Task 14 corpus was used for evaluating the various fusion methods for turning word into sentence ratings. All unseen words in the SemEval corpus are added to the lexicon using the affective lexicon expansion algorithm outlined above. The model used to create the required ratings is trained using all of the words in the ANEW corpus as training samples and 300 of them as seed words, i.e., $N = 300$ for this experiment. Then the ratings of the words are combined to create the sentence rating using linear fusion (3), weighted average fusion (4) or non-linear max fusion (5). In the first experiment (labeled "all words"), all words in a sentence are taken into account to compute the sentence score. In the second experiment (labeled "content words"), only the verbs, nouns, adjectives and adverbs are used.

Similarly for the subtitles corpus, all required word ratings were created by a model trained with ANEW data, then combined into sentence ratings. The temporal dependencies of the subtitles are disregarded (sentences occurring close to each other are likely to have similar content), each sentence is handled out of context. For this experiment, all words in the subtitles are used.

In order to adapt the affective model to the ChIMP task, the discrete sentence level valence scores were mapped as follows: frustrated was assigned a valence value of -1, neutral was 0 and polite was 1. To bootstrap the valence scores for each word in the ChIMP corpus, we used the average sentence-level scores for all sentence where that word appeared. Finally, the ANEW equation system matrix was augmented with all the words in the ChIMP corpus and the valence model in (2) was estimated using LMS. Note that for this training process a 10-fold cross validation experiment was run on the ChIMP corpus sentences. The relative weight of the ChIMP corpus adaptation data was varied by adding the respective lines multiple times to the augmented system matrix, e.g., adding each line twice gives a weight of $w = 2$. We tested weights of $w = 1$, $w = 2$, and using only the samples from ChIMP as training samples (denoted as $w = \infty$). The valence boundary between frustrated and other classes was selected based on the a-priori probability distribution for each class, and is simply the Bayesian

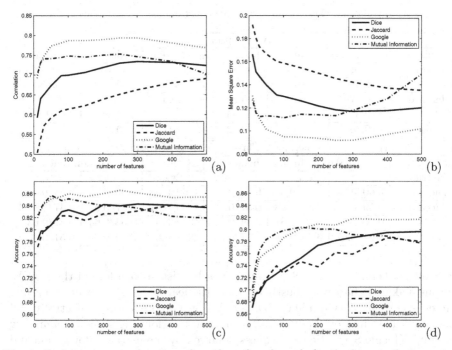

Fig. 1. Performance as a function of number of seed words for the four similarity metrics: (a) correlation between the automatically computed and manually annotated scores (ANEW corpus), (b) mean square error (ANEW), (c) binary (positive vs negative) classification accuracy (ANEW), (d) binary classification accuracy (General Inquirer)

decision boundary (similarly between polite and other classes). The focus of this experiment is the effectiveness of the adaptation procedure, therefore only results using Google Semantic Relatedness and all words of each sentence are presented.

In order to evaluate the performance of the sentence level affective scores we use the following metrics: (i) Pearson correlation between the manually labeled and automatically computed scores, (ii) classification accuracy for the 2-class (positive, negative) and 3-class (positive, neutral, negative) problems, and (iii) precision, recall and F-score for the 2-class (positive vs negative) problem.

4 Results for Affective Lexicon Creation

Figure 1 shows the performance of the affective lexicon creation algorithm on the ANEW (a)-(c) and General Inquirer corpora (d) as a function of the number of features N (seed words) and for each semantic similarity metric. Overall, the results obtained are very good both in terms of correlation and binary classification accuracy. Performance improves up to about $N = 200$ to 300 seed words and then levels off or falls slightly. Note, that good performance is achieved even with few features (less than 100 features) especially when using the mutual information or Google semantic relatedness metrics. Although, more features should

Table 2. Correlation, class. accuracy for SemEval dataset wrt metrics and fusion.

Correlation						
Similarity	All Words: fusion scheme			Content Words: fusion scheme		
Metric	avg	w.avg	max	avg	w.avg	max
Dice	0.48	0.46	0.44	0.5	0.47	0.45
Jaccard	0.37	0.3	0.28	0.45	0.39	0.36
PMI	0.37	0.31	0.25	0.4	0.34	0.27
Google	0.44	0.44	0.42	0.47	0.46	0.44
3-Class (pos., neutral, neg.) / 2-Class (pos., neg.) Classification Accuracy						
Similarity	All Words: fusion scheme			Content Words: fusion scheme		
Metric	avg	w.avg	max	avg	w.avg	max
Dice	0.60/0.65	0.59/0.68	0.53/0.70	0.60/0.68	0.60/0.69	0.54/0.71
Jaccard	0.59/0.59	0.42/0.58	0.31/0.58	0.60/0.67	0.57/0.68	0.47/0.67
PMI	0.60/0.60	0.54/0.61	0.34/0.61	0.59/0.64	0.54/0.64	0.37/0.63
Google	0.60/0.67	0.59/0.68	0.51/0.69	0.60/0.69	0.59/0.69	0.51/0.69

in principle help, only $K = 827$ training words exist in each fold of the ANEW corpus making it hard to robustly estimate the feature weights (over-fitting). A larger seed vocabulary would enable us to use even more features effectively and possibly lead to improved performance . The best performing semantic similarity metric is Google Semantic Relatedness, followed by the mutual information and Dice metrics. This trend is consistent in all experiments shown in Fig. 1. Note that PMI is not very well suited to this task, since it is the only one among the metrics that is unbounded. This is a possible explanation for the different performance dynamics it exhibits, peaking much earlier (around $N = 50$ to 150) and then trailing off. However, PMI it still one of the top performers. In terms of absolute numbers, the results reported here are at least as good as the state-of-the-art. Correlation results for the ANEW corpus shown in Fig. 1(a) are between 0.72 and 0.78. Binary classification results for the ANEW corpus shown in Fig. 1(c) are between 82% and 86% for the various metrics. Binary classification accuracy for the General Inquirer corpus shown in Fig. 1(d) is up to 82% using Google Semantic Relatedness. Compare this to 82.8% accuracy quoted in [20] and [21] using NEAR queries for computing semantic similarity, 82.1% in [6], and 81.9% in [19]. Note that the latter two methods achieve higher performance when using part of the GI corpus itself as training. Overall, the results are impressive given that ANEW ratings were used to seed the GI corpus and the differences in the manual tagging procedure for each corpus.

5   Results for Sentence Level Tagging

SemEval Dataset: The performance for the SemEval sentence level affective tagging is summarized in Table 2. Results are shown for 3-way and 2-way classification experiments.

For 3-class experiments the following mapping from continuous to discrete values was performed: $[-1, -0.5) \rightarrow -1$ (negative), $[-0.5, 0.5) \rightarrow 0$ (neutral), and $[0.5, 1] \to 1$ (positive). Similarly for 2-class: $[-1, 0) \to -1$ (negative) and $[0, 1] \to 1$ (positive). The correlation and 3-way accuracy numbers are higher than those reported in [16] (47.7% and 55.1% respectively) and 2-way accuracy is comparable to that achieved by [12] (71%). The relative performance of the various similarity metrics is different than for the word rating task. Google semantic relatedness performed best at the word level, but is the second-best performer at the sentence level. Conversely, Dice that was an average performer on words, works the best on sentences. When comparing the three fusion schemes, linear fusion performs well in terms of correlation, 2-way and 3-way classification. Non-linear fusion schemes (weighted average and, especially, max) work best for the binary classification task, giving more weight to words with extreme valence scores. Finally, using only content words (instead of all words) improves results, yet the gain is minimal in almost all cases. This indicates that non-content words already have been assigned low word rating scores. Overall, the results are good, significantly higher than those reported in [18] and on par with the best systems reported in [16] and [12], when evaluating performance on *all* the sentences in the dataset. Best results in terms of correlation of 0.5 are achieved using linear fusion. 3-way classification accuracy (although higher than any system reported in [16]) is still poor, since the a-priori probability for the neutral class is 0.6 and our best performance, achieved using linear (or weighted average) fusion, barely matched that at 60%. Finally, non-linear max fusion achieves the best results for 2-way classification at 71%.

Subtitles Dataset: Table 3 shows performance in the subtitles dataset. The results are fairly low, with accuracy under 60%, and very small correlation. The relatively poor results show the added complexity of this task compared to uni-modal polarity detection in text. More likely it points to the significance of factors we ignored: interactions across sentences and across modalities. It is reasonable to assume that context acts as a modifier on the affective interpretation of each utterance. Furthermore, here, it is not just lexical context that contributes to the ground truth, but also multimedia context: from the voice tone of the actor, to his facial expression, to the movie's setting.

ChIMP Corpus: In Table 4, the two-class sentence-level classification accuracy is shown for the ChIMP corpus (polite vs other: "P vs O", frustrated vs other: "F vs O"). For the adaptation experiments on the ChIMP corpus, the parameter w denotes the weighting given to the in-domain ChIMP data, i.e., number of

Table 3. Performance on subtitles dataset for Google metric wrt fusion

Performance	Fusion scheme		
Measurement	avg	w.avg	max
Correlation	0.05	0.05	0.06
2-Class Classification Accuracy (positive, negative)	0.58	0.56	0.56
Precision (positive vs negative)	0.62	0.62	0.61
Recall (positive vs negative	0.85	0.84	0.82
F-score (positive vs negative)	0.72	0.71	0.70

Table 4. Sentence classification accuracy for ChIMP baseline and adapted tasks

Sentence Classification	Fusion scheme		
Accuracy	avg	w.avg	max
Baseline: P vs O / F vs O	0.70/0.53	0.69/0.62	0.54/0.66
Adapt $w = 1$: P vs O / F vs O	0.74/0.51	0.70/0.58	0.67/0.57
Adapt $w = 2$: P vs O / F vs O	0.77/0.49	0.74/0.53	0.71/0.53
Adapt $w = \infty$: P vs O / F vs O	0.84/0.52	0.82/0.52	0.75/0.52

times the adaptation equations were repeated in the system matrix (2). Results are shown for the three fusion methods (average, weighted average, maximum). For the ChIMP politeness detection task, performance of the baseline (unsupervised) model is lower than that quoted in [22] for lexical features. Performance improves significantly by adapting the affective model using in-domain ChIMP data reaching up to 84% accuracy for linear fusion (matching the results in [22]). The best results for frustration detection are achieved with the baseline model and max fusion schemes at 66% (at least as good as those reported in [22]).

6   Conclusions

We proposed an affective lexicon creation/expansion algorithm that estimates a continuous valence score for unseen words from a set of manually labeled seed words and semantic similarity ratings. The lexicon creation algorithm achieved very good results on the ANEW and General Inquirer datasets using 200-300 seed words, achieving correlation scores of up to 0.79 and over 80% binary classification accuracy. In addition, preliminary results on sentence level valence estimation show that simple fusion schemes achieve performance that is at least at a par with the state-of-the-art for the SemEval task. For politeness detection it was shown that adaptation of the affective model and linear fusion achieves the best results. For frustration detection, the domain-independent model and max fusion gave the best performance. Polarity detection on the movie subtitles dataset proved the most challenging, showing the importance of context and multimodal information for estimating affect. Overall, we have shown that an unsupervised domain-independent approach is a viable alternative to training domain-specific language models for the problem of affective text analysis.

Although the results are encouraging, this work represents only the first step towards applying machine learning methods to affective text analysis. Future research direction could involve more complex models (e.g., non-linear, kernel) to model lexical affect, alternative semantic similarity or relatedness metrics, and better fusion models that incorporate syntactic and pragmatic information to improve sentence-level valence scores. Overall, the proposed method creates reasonably accurate ratings and is a good starting point for future research. An important advantage of the proposed method is it's simplicity: the only requirements for constructing EmotiWord are a few hundred labeled seed words and a web search engine.

References

1. Andreevskaia, A., Bergler, S.: CLaC and CLaC-NB: Knowledge-based and corpus-based approaches to sentiment tagging. In: Proc. SemEval, pp. 117–120 (2007)
2. Ang, J., Dhillon, R., Krupski, A., Shriberg, E., Stolcke, A.: Prosody-based automatic detection of annoyance and frustration in human-computer dialog. In: Proc. ICSLP, pp. 2037–2040 (2002)
3. Bradley, M., Lang, P.: Affective norms for english words (ANEW): Stimuli, instruction manual and affective ratings. Technical report C-1. The Center for Research in Psychophysiology, University of Florida (1999)
4. Chaumartin, F.R.: UPAR7: A knowledge-based system for headline sentiment tagging. In: Proc. SemEval, pp. 422–425 (2007)
5. Esuli, A., Sebastiani, F.: Sentiwordnet: A publicly available lexical resource for opinion mining. In: Proc. LREC, pp. 417–422 (2006)
6. Hassan, A., Radev, D.: Identifying text polarity using random walks. In: Proc. ACL, pp. 395–403 (2010)
7. Hatzivassiloglou, V., McKeown, K.: Predicting the Semantic Orientation of Adjectives. In: Proc. ACL, pp. 174–181 (1997)
8. Iosif, E., Potamianos, A.: Unsupervised Semantic Similarity Computation Between Terms Using Web Documents. IEEE Transactions on Knowledge and Data Engineering 22(11), 1637–1647 (2009)
9. Lee, C.M., Narayanan, S.S., Pieraccini, R.: Combining acoustic and language information for emotion. In: Proc. ICSLP, pp. 873–876 (2002)
10. Lee, C.M., Narayanan, S.S.: Toward detecting emotions in spoken dialogs. IEEE Transactions on Speech and Audio Processing 13(2), 293–303 (2005)
11. Malandrakis, N., Potamianos, A., Evangelopoulos, G., Zlatintsi, A.: A supervised approach to movie emotion tracking. In: Proc. ICASSP, pp. 2376–2379 (2011)
12. Moilanen, K., Pulman, S., Zhang, Y.: Packed feelings and ordered sentiments: Sentiment parsing with quasi-compositional polarity sequencing and compression. In: Proc. WASSA Workshop at ECAI, pp. 36–43 (2010)
13. Pelletier, F.J.: The principle of semantic compositionality. Topoi 13, 11–24 (1994)
14. Polanyi, L., Zaenen, A.: Contextual valence shifters. In: Computing attitude and affect in text: Theory and Applications, pp. 1–10. Springer (2006)
15. Stone, P., Dunphy, D., Smith, M., Ogilvie, D.: The General Inquirer: A Computer Approach to Content Analysis. The MIT Press (1966)
16. Strapparava, C., Mihalcea, R.: Semeval-2007 task 14: Affective text. In: Proc. SemEval, pp. 70–74 (2007)
17. Strapparava, C., Valitutti, A.: WordNet-Affect: an affective extension of WordNet. In: Proc. LREC, vol. 4, pp. 1083–1086 (2004)
18. Taboada, M., Brooke, J., Tofiloski, M., Voll, K., Stede, M.: Lexicon-based methods for sentiment analysis. Computational Linguistics 1, 1–41 (2010)
19. Takamura, H., Inui, T., Okumura, M.: Extracting semantic orientations of words using spin model. In: Proc ACL, pp. 133–140 (2005)
20. Turney, P., Littman, M.L.: Unsupervised Learning of Semantic Orientation from a Hundred-Billion-Word Corpus. Technical report ERC-1094 (NRC 44929). National Research Council of Canada (2002)
21. Turney, P., Littman, M.L.: ACM Trans. on Information Systems. Measuring praise and criticism: Inference of semantic orientation from association 21, 315–346 (2003)
22. Yildirim, S., Narayanan, S., Potamianos, A.: Detecting emotional state of a child in a conversational computer game. Computer Speech and Language 25, 29–44 (2011)

A Bayesian Active Learning Framework for a Two-Class Classification Problem

Pablo Ruiz[1], Javier Mateos[1], Rafael Molina[1],
and Aggelos K. Katsaggelos[2]

[1] University of Granada, 18071 Granada, Spain
mataran@decsai.ugr.es
http://decsai.ugr.es/vip
[2] Northwestern University, Evanston, IL, USA

Abstract. In this paper we present an active learning procedure for the two-class supervised classification problem. The utilized methodology exploits the Bayesian modeling and inference paradigm to tackle the problem of kernel-based data classification. This Bayesian methodology is appropriate for both finite and infinite dimensional feature spaces. Parameters are estimated, using the kernel trick, following the evidence Bayesian approach from the marginal distribution of the observations. The proposed active learning procedure uses a criterion based on the entropy of the posterior distribution of the adaptive parameters to select the sample to be included in the training set. A synthetic dataset as well as a real remote sensing classification problem are used to validate the followed approach.

1 Introduction

In many real applications large collections of data are extracted whose class is unknown. Those applications include, for instance, most image classification applications, text processing, speech recognition, and biological research problems. While extracting the samples is straightforward and inexpensive, classifying each one of those samples is a tedious and often expensive task. Active learning is a supervised learning technique that attempts to overcome the labeling bottleneck by asking queries in the form of unlabeled samples to be labeled by an *oracle* (e.g., a human annotator) [10]. An active learning procedure queries only the most informative samples from the whole set of unlabeled samples. The objective is to obtain a high classification performance using as few labeled samples as possible, minimizing, this way, the cost of obtaining labeled data.

Kernel methods in general and Support Vector Machines (SVMs) in particular dominate the field of discriminative data classification [8]. This problem has also been approached from a Bayesian point of view. For example, the relevance vector machine [13] assumes a Gaussian prior over the adaptive parameters and uses the EM algorithm to estimate them. In practice, this prior enforces sparsity because the posterior distribution of many adaptive parameters is sharply peaked around zero. Lately, Gaussian Process Classification [7] has received much attention. Adopting the least-squares SVM formulation may alternatively allow to

E. Salerno, A.E. Çetin, and O. Salvetti (Eds.): MUSCLE 2011, LNCS 7252, pp. 42–53, 2012.

perform Bayesian inference on SVMs [12]. A huge benefit is obtained by applying Bayesian inference on these machines since hyperparameters may be learned directly from data using a consistent theoretical framework.

In this paper we make use of the Bayesian paradigm to tackle the problem of active learning on kernel-based two-class data classification. The Bayesian modeling and inference approach to the kernel-based classification we propose in this paper allows us to derive efficient closed-form expressions for parameter estimation and active learning.

The general two-class supervised classification problem [2] we tackle here implies a classification function of the form:

$$y(\mathbf{x}) = \boldsymbol{\phi}^\top(\mathbf{x})\mathbf{w} + b + \epsilon, \tag{1}$$

where the mapping $\boldsymbol{\phi} : \mathcal{X} \to \mathcal{H}$ embeds the observed $\mathbf{x} \in \mathcal{X}$ into a higher L-dimensional (possibly infinite) feature space $\mathcal{H}$. The output $y(\mathbf{x}) \in \{0, 1\}$ consists of a binary coding representation of its classification, $\mathbf{w}$ is a vector of size $L \times 1$ of adaptive parameters to be estimated, b represents the bias in the classification function, and ϵ is an independent realization of the Gaussian distributions $\mathcal{N}(0, \sigma^2)$.

While kernel-based classification in *static* scenarios has been extensively studied, the problem related to the emerging field of *active learning* [10] is still unsolved. Let us assume that we have access to P vectors in the feature space denoted by $\boldsymbol{\phi}(\mathbf{x}_i), i = 1, \ldots, P$ for which the corresponding output $y(\mathbf{x}_i), i = 1, \ldots, P$ can be provided by an oracle. The key is to decide which elements $\mathbf{x}_i$ to acquire from the set of P possible samples in order to build an optimal compact classifier. Active learning aims at efficiently sampling the observations space to improve the model performance by *incrementally* building training sets. Such sets are obtained by selecting from the available samples the best ones according to a selection strategy and querying the oracle only for the label of those samples. Many selection strategies have been devised in the literature, which are based on different heuristics: 1) large margin, 2) expert committee, and 3) posterior probability (see [10] for a comprehensive review). The first two approaches typically exploit SVM methods. The latter requires classifiers that can provide posterior probabilities.

In [6], a Bayesian active learning procedure for finite dimensional feature spaces is proposed. Assuming that $\boldsymbol{\phi}(\mathbf{x}_i), i = 1, \ldots, P$ has L components, the design matrix $\boldsymbol{\Phi}_{:,:}$ is of size $P \times L$, whose i^{th} row, $i = 1, \ldots, P$ is given by $\boldsymbol{\phi}(\mathbf{x}_i)^\top$. Then, a subset of size C of the L columns of $\boldsymbol{\Phi}_{:,:}$, denoted by $\boldsymbol{\Phi}_{:,I_C}$, is selected using the differential entropy instead of the response functions $y(\mathbf{x}_i)$ [6]. Notice that this approach is in contrast to other basis selection techniques which make explicit use of the response functions, for example, [3] in the context of SVM, [4] in the context of sparse representation, and [1] considering compressive sensing. To select the rows of $\boldsymbol{\Phi}_{:,I_C}$, for which the response associated to $\boldsymbol{\phi}(\mathbf{x}_i)$ will be queried, a criterion based again on differential entropy is utilized (see [6] for details). See also [5] for the general theory and [9] for the use of the approach in compressive sensing.

Here, the Bayesian modeling and inference paradigm is applied to two-class classification problems which utilize kernel-based classifiers. This paradigm is used to tackle both active learning and parameter estimation for infinite dimensional feature spaces, and consequently for problems where basis selection cannot be carried out explicitly. As we will see later, the proposed approach will make extensive use of the marginal distribution of the observations to avoid dealing with infinite dimensional feature spaces and the posterior distribution of the infinite dimensional $\mathbf{w}$.

The rest of the paper is organized as follows. Section 2 introduces the models we use in our Bayesian framework. Then, in section 3, Bayesian inference is performed. We calculate the posterior distribution of $\mathbf{w}$, and propose a methodology for parameter estimation, active learning, and class prediction. Experiments illustrating the performance of the proposed approach on a synthetic and a real remote sensing classification problem are presented in section 4. Finally, section 5 concludes the paper.

2   Bayesian Modeling

Let us assume that the target variable $y(\mathbf{x}_i)$ follows the model in Eq. (1). If we already know the classification output $y(\mathbf{x}_i)$ associated with the feature samples $\phi(\mathbf{x}_i), i = 1, \ldots, M$, with M the number of samples, we can then write

$$p(\mathbf{y}|\mathbf{w}, \sigma^2) = \prod_{i=1}^{M} \mathcal{N}(y(\mathbf{x}_i)|\phi^{\top}(\mathbf{x}_i)\mathbf{w} + b, \sigma^2). \tag{2}$$

Since $\mathbf{x}_i, i = 1, \ldots, M$, will always appear as conditioning variable, for the sake of simplicity, we have removed the dependency on $\mathbf{x}_1, \ldots, \mathbf{x}_M$ in the left-hand side of the equation. We note that, for infinite dimensional feature vectors $\phi(\mathbf{x}_i)$, $\mathbf{w}$ is infinite dimensional.

The Bayesian framework allows us to introduce information about the possible value of $\mathbf{w}$ in the form of a prior distribution. In this work we assume that each component of $\mathbf{w}$ independently follows a Gaussian distribution $\mathcal{N}(0, \gamma^2)$. When the feature vectors are infinite dimensional, we will not make explicit use of this prior distribution but still we will be able to carry out parameter estimation and active learning tasks.

3   Bayesian Inference

Bayesian inference extracts conclusions from the posterior distribution $p(\mathbf{w}|\mathbf{y}, \gamma^2, \sigma^2)$. The posterior distribution of $\mathbf{w}$ is given by [2]

$$p(\mathbf{w}|\mathbf{y}, \gamma^2, \sigma^2) = \mathcal{N}(\mathbf{w}|\boldsymbol{\Sigma}_{\mathbf{w}|\mathbf{y}, \gamma^2, \sigma^2}\sigma^{-2}\boldsymbol{\Phi}^{\top}(\mathbf{y} - b\mathbf{1}), \boldsymbol{\Sigma}_{\mathbf{w}|\mathbf{y}, \gamma^2, \sigma^2}), \tag{3}$$

where

$$\boldsymbol{\Sigma}_{\mathbf{w}|\mathbf{y}, \gamma^2, \sigma^2} = (\sigma^{-2}\boldsymbol{\Phi}^{\top}\boldsymbol{\Phi} + \gamma^{-2}\mathbf{I})^{-1}$$

and $\boldsymbol{\Phi}$ is the design matrix whose i^{th} row is $\phi(\mathbf{x}_i)^{\top}$.

It is important to note that we do not need to know the form of $\boldsymbol{\Phi}$ explicitly to calculate this posterior distribution. We only need to know the Gram matrix $\mathbf{K} = \boldsymbol{\Phi}\boldsymbol{\Phi}^\top$, which is an $M \times M$ symmetric matrix with elements $\mathbf{K}_{nm} = k(\mathbf{x}_n, \mathbf{x}_m) = \phi(\mathbf{x}_n)^\top \phi(\mathbf{x}_m)$, which has to be a positive semidefinite matrix [8]. This leads to the construction of kernel functions $k(\mathbf{x}, \mathbf{x}')$ for which the Gram matrix $\mathbf{K}$ is positive semidefinite for all possible choices of the set $\{\mathbf{x}_n\}$ [11]. Note that, even if $\boldsymbol{\Phi}$ has an infinite number of columns, which correspond to the case of $\mathbf{x}_i$ being an infinite dimensional feature vector, we can still calculate $\mathbf{K}$ of size $M \times M$ by means of the kernel function. Note also that we are somewhat abusing the notation here because $\mathbf{w}$ is infinite dimensional for infinite dimensional feature vectors.

3.1 Parameter Estimation

To estimate the values of γ^2 and σ^2 we use the Evidence Bayesian approach without any prior information on these parameters. According to it, we maximize the marginal distribution obtained by integrating out the vector of adaptive parameters $\mathbf{w}$. It can easily be shown, see for instance [2], that

$$p(\mathbf{y}|\gamma^2, \sigma^2) = \mathcal{N}(\mathbf{y}|b\mathbf{1}, \boldsymbol{\Sigma}_{\mathbf{y}|\gamma^2,\sigma^2}), \tag{4}$$

where

$$\boldsymbol{\Sigma}_{\mathbf{y}|\gamma^2,\sigma^2} = \gamma^2\boldsymbol{\Phi}\boldsymbol{\Phi}^\top + \sigma^2\mathbf{I}.$$

The value of b can be easily obtained from Eq. (4) as

$$b = \frac{1}{M}\sum_{i=1}^M y(\mathbf{x}_i). \tag{5}$$

Differentiating $2\ln p(\mathbf{y}|\gamma^2, \sigma^2)$ with respect to γ^2 and equating to zero, we obtain

$$\begin{aligned}
&\mathbf{tr}[(\gamma^2\boldsymbol{\Phi}\boldsymbol{\Phi}^\top + \sigma^2)^{-1}\boldsymbol{\Phi}\boldsymbol{\Phi}^\top] = \\
&\mathbf{tr}[(\mathbf{y} - b\mathbf{1})^\top(\gamma^2\boldsymbol{\Phi}\boldsymbol{\Phi}^\top + \sigma^2\mathbf{I})^{-1}\boldsymbol{\Phi}\boldsymbol{\Phi}^\top(\gamma^2\boldsymbol{\Phi}\boldsymbol{\Phi}^\top + \sigma^2\mathbf{I})^{-1}(\mathbf{y} - b\mathbf{1})].
\end{aligned} \tag{6}$$

Diagonalizing $\boldsymbol{\Phi}\boldsymbol{\Phi}^\top$, we obtain $\mathbf{U}\boldsymbol{\Phi}\boldsymbol{\Phi}^\top\mathbf{U}^\top = \mathbf{D}$, where $\mathbf{U}$ is an orthonormal matrix and $\mathbf{D}$ is a diagonal matrix with entries $\lambda_i, i = 1, \ldots, M$. We can then rewrite the above equation as

$$\sum_{k=1}^M \frac{\lambda_k}{\gamma^2\lambda_k + \sigma^2} = \sum_{i=1}^M z_i^2 \frac{\lambda_i}{(\gamma^2\lambda_i + \sigma^2)^2}, \tag{7}$$

where $\mathbf{U}(\mathbf{y} - b\mathbf{1}) = \mathbf{z}$ with components $z_i, i = 1, \ldots, M$.

Multiplying both sides of the above equation by γ^2 we have

$$\gamma^2 = \sum_{i=1}^M \frac{\frac{\lambda_i}{\gamma^2\lambda_i + \sigma^2}}{\sum_{k=1}^M \frac{\lambda_k}{\gamma^2\lambda_k + \sigma^2}} \frac{\gamma^2 z_i^2}{\gamma^2\lambda_i + \sigma^2} = \sum_{i=1}^M \mu_i \frac{\gamma^2 z_i^2}{\gamma^2\lambda_i + \sigma^2}, \tag{8}$$

where

$$\mu_i = \frac{\frac{\lambda_i}{\gamma^2 \lambda_i + \sigma^2}}{\sum_{k=1}^{M} \frac{\lambda_k}{\gamma^2 \lambda_k + \sigma^2}}. \tag{9}$$

Note that $\mu_i \geq 0$ and $\sum_{i=1}^{M} \mu_i = 1$.

Similarly, differentiating $2 \ln p(\mathbf{y}|\gamma^2, \sigma^2)$ with respect to σ^2 and equating it to zero, we obtain

$$\sum_{k=1}^{M} \frac{1}{\gamma^2 \lambda_k + \sigma^2} = \sum_{i=1}^{M} z_i^2 \frac{1}{(\gamma^2 \lambda_i + \sigma^2)^2}. \tag{10}$$

Following the same steps we already performed to estimate γ^2, we obtain

$$\sigma^2 = \sum_{i=1}^{M} \nu_i \frac{\sigma^2 z_i^2}{\gamma^2 \lambda_i + \sigma^2}, \tag{11}$$

where

$$\nu_i = \frac{\frac{1}{\gamma^2 \lambda_i + \sigma^2}}{\sum_{k=1}^{M} \frac{1}{\gamma^2 \lambda_k + \sigma^2}}. \tag{12}$$

Note that, again, $\nu_i \geq 0$ and $\sum_{i=1}^{M} \nu_i = 1$.

To obtain estimates of γ^2 and σ^2 we use an iterative procedure where the values of the old estimates of γ^2 and σ^2 are used on the right hand side of Equations (8) and (11) to obtain the updated values of the parameters in the left hand side of these equations. Although we have not formally established the convergence and unicity of the solution, we have not observed any convergence problems in the performed experiments. Note that to estimate γ^2 and σ^2 we have not made use of the posterior distribution of the components of $\mathbf{w}$.

3.2   Active Learning

Active learning starts with a small set of observations whose class is already known. From these observations, the posterior distribution of $\mathbf{w}$ and the parameters b, γ^2 and σ^2 can be estimated using the procedure described in the previous sections. Now we want that the system learns new observations incrementally. Let us assume that we want to add a new observation associated to $\phi(\mathbf{x}_+)$, whose corresponding $y(\mathbf{x}_+)$ will be learned by querying the oracle. The covariance matrix of the posterior distribution of $\mathbf{w}$ when $\phi(\mathbf{x}_+)$ is added is given by

$$\Sigma_{\mathbf{w}|\mathbf{y},\gamma^2\sigma^2}^{\mathbf{x}_+} = (\sigma^{-2}(\boldsymbol{\Phi}^\top\boldsymbol{\Phi} + \phi(\mathbf{x}_+)\phi^\top(\mathbf{x}_+)) + \gamma^{-2}\mathbf{I})^{-1}.$$

Since we have a set of observations that could be added and whose class is unknown (but can be learned by querying the oracle), the objective of active learning is to select the observation that maximizes the performance of the system, minimizing in this way the number of queries answered by the oracle. To select this new feature vector, in this paper, we propose to maximize the difference between the entropies of the posterior distribution before and after adding the new feature vector (see [6, 9]) to obtain

$$\mathbf{x}_+ = \arg\max_{\mathbf{x}} \frac{1}{2} \log |\boldsymbol{\Sigma}_{\mathbf{w}|\mathbf{y},\gamma^2,\sigma^2}||\boldsymbol{\Sigma}^{\mathbf{x}}_{\mathbf{w}|\mathbf{y},\gamma^2,\sigma^2}|^{-1}. \tag{13}$$

Then we have

$$\frac{1}{2} \log |\boldsymbol{\Sigma}_{\mathbf{w}|\mathbf{y},\gamma^2,\sigma^2}||\boldsymbol{\Sigma}^{\mathbf{x}}_{\mathbf{w}|\mathbf{y},\gamma^2,\sigma^2}|^{-1}$$

$$= \frac{1}{2} \log |\mathbf{I} + \sigma^{-2}\boldsymbol{\phi}(\mathbf{x})\boldsymbol{\phi}^\top(\mathbf{x})(\sigma^{-2}\boldsymbol{\Phi}^\top\boldsymbol{\Phi} + \gamma^{-2}\mathbf{I})^{-1}|$$

$$= \frac{1}{2} \log(1 + \sigma^{-2}\boldsymbol{\phi}^\top(\mathbf{x})(\sigma^{-2}\boldsymbol{\Phi}^\top\boldsymbol{\Phi} + \gamma^{-2}\mathbf{I})^{-1}\boldsymbol{\phi}(\mathbf{x})),$$

and using

$$(\sigma^{-2}\boldsymbol{\Phi}^\top\boldsymbol{\Phi} + \gamma^{-2}\mathbf{I})^{-1} = \gamma^2\mathbf{I} - \gamma^4\boldsymbol{\Phi}^\top(\sigma^2\mathbf{I} + \gamma^2\boldsymbol{\Phi}\boldsymbol{\Phi}^\top)^{-1}\boldsymbol{\Phi}, \tag{14}$$

we can finally write

$$\frac{1}{2} \log |\boldsymbol{\Sigma}_{\mathbf{w}|\mathbf{y},\gamma^2,\sigma^2}| \cdot |\boldsymbol{\Sigma}^{\mathbf{x}}_{\mathbf{w}|\mathbf{y},\gamma^2,\sigma^2}|^{-1}$$

$$= \frac{1}{2} \log \left(1 + \sigma^{-2}\gamma^2\boldsymbol{\phi}^\top(\mathbf{x})\boldsymbol{\phi}(\mathbf{x}) - \sigma^{-2}\gamma^4\boldsymbol{\phi}^\top(\mathbf{x})\boldsymbol{\Phi}^\top(\sigma^2\mathbf{I} + \gamma^2\boldsymbol{\Phi}\boldsymbol{\Phi}^\top)^{-1}\boldsymbol{\Phi}\boldsymbol{\phi}(\mathbf{x})\right)$$

$$= \frac{1}{2} \log \left(1 + \sigma^{-2}\gamma^2\boldsymbol{\phi}^\top(\mathbf{x})\boldsymbol{\phi}(\mathbf{x}) - \sigma^{-2}\gamma^4\boldsymbol{\phi}^\top(\mathbf{x})\boldsymbol{\Phi}^\top \boldsymbol{\Sigma}^{-1}_{\mathbf{y}|\gamma^2,\sigma^2}\boldsymbol{\Phi}\boldsymbol{\phi}(\mathbf{x})\right). \tag{15}$$

Consequently, all needed quantities to select $\mathbf{x}_+$ can be calculated without knowledge of the feature vectors and the posterior distribution of the possibly infinite dimensional adaptive parameters and using only kernel functions and the marginal distribution of the observations.

Notice that, given $\boldsymbol{\Sigma}^{-1}_{\mathbf{y}|\gamma^2,\sigma^2}$, we can easily calculate the new precision matrix $\boldsymbol{\Sigma}^{-1}_{\mathbf{y},y(\mathbf{x}_+)|\gamma^2,\sigma^2}$ of the marginal distribution of $\mathbf{y}$ when the observation corresponding to $\mathbf{x}_+$ has been added. We have

$$\boldsymbol{\Sigma}^{-1}_{\mathbf{y},y(\mathbf{x}_+)|,\gamma^2,\sigma^2} = \begin{pmatrix} \mathbf{M} & -\mathbf{M}\mathbf{v}d^{-1} \\ -d^{-1}\mathbf{v}^\top\mathbf{M} & d^{-1} + d^{-2}\mathbf{v}^\top\mathbf{M}\mathbf{v} \end{pmatrix}, \tag{16}$$

with $\mathbf{v} = \gamma^2\boldsymbol{\Phi}\boldsymbol{\phi}(\mathbf{x}_+)$, $d = \sigma^2 + \gamma^2\boldsymbol{\phi}^\top(\mathbf{x}_+)\boldsymbol{\phi}(\mathbf{x}_+)$, and $\mathbf{M} = (\boldsymbol{\Sigma}_{\mathbf{y}|\gamma^2,\sigma^2} - d^{-1}\mathbf{v}\mathbf{v}^\top)^{-1}$. To calculate $\mathbf{M}$ we use the Sherman-Morrison-Woodbury formula to obtain

$$\mathbf{M} = \boldsymbol{\Sigma}^{-1}_{\mathbf{y}|\gamma^2,\sigma^2} - \frac{1}{-d + \mathbf{v}^\top\boldsymbol{\Sigma}^{-1}_{\mathbf{y}|\gamma^2,\sigma^2}\mathbf{v}}\boldsymbol{\Sigma}^{-1}_{\mathbf{y}|\gamma^2,\sigma^2}\mathbf{v}\mathbf{v}^\top\boldsymbol{\Sigma}^{-1}_{\mathbf{y}|\gamma^2,\sigma^2},$$

and consequently $\boldsymbol{\Sigma}^{-1}_{\mathbf{y},y(\mathbf{x}_+)|\gamma^2,\sigma^2}$ can be calculated from the previous $\boldsymbol{\Sigma}^{-1}_{\mathbf{y}|\gamma^2,\sigma^2}$ in a straightforward manner.

Hence, starting with an initial estimation of the parameters, to perform active learning we alternate between the selection of a new sample using Eq. (13) and the estimation of the unknown parameters b, γ^2, and σ^2 using the procedure described in section 3.1.

3.3   Prediction

Once the system has been trained, we want to predict the value of $y(\mathbf{x}_*)$ for a new value of $\mathbf{x}$, denoted by $\mathbf{x}_*$. To calculate this predicted value, we make use of the distribution of $\boldsymbol{\phi}^\top(\mathbf{x}_*)\mathbf{w} + b$ where the posterior distribution of $\mathbf{w}$ is given in Eq. (3). Its mean value, $\boldsymbol{\phi}^\top(\mathbf{x}_*)\mathrm{E}[\mathbf{w}] + b$, is given by

$$\boldsymbol{\phi}^\top(\mathbf{x}_*)\mathrm{E}[\mathbf{w}] + b = \boldsymbol{\phi}^\top(\mathbf{x}_*)\boldsymbol{\Sigma}_{\mathbf{w}|\mathbf{y},\gamma^2,\sigma^2}\sigma^{-2}\boldsymbol{\Phi}^\top(\mathbf{y} - b\mathbf{1}) + b, \qquad (17)$$

where we have made use of Eq. (14) to obtain

$$\begin{aligned}
\boldsymbol{\phi}^\top(\mathbf{x}_*)\mathrm{E}[\mathbf{w}] + b = {} & \gamma^2\sigma^{-2}\boldsymbol{\phi}^\top(\mathbf{x}_*)\boldsymbol{\Phi}^\top(\mathbf{y} - b\mathbf{1}) \qquad\qquad (18)\\
& - \gamma^4\sigma^{-2}\boldsymbol{\phi}^\top(\mathbf{x}_*)\boldsymbol{\Phi}^\top(\sigma^2\mathbf{I} + \gamma^2\boldsymbol{\Phi}\boldsymbol{\Phi}^\top)^{-1}\boldsymbol{\Phi}\boldsymbol{\Phi}^\top(\mathbf{y} - b\mathbf{1}) + b,
\end{aligned}$$

which can be calculated without knowing the feature vectors if the kernel function is known.

4   Experimental Results

We have tested the proposed active learning algorithm on a synthetic dataset and a real remote sensing classification problem. The synthetic data set, due to Paisley [6], consists of 200 observations, 100 from each one of the two classes, in a bi-dimensional space. The data, plotted in figure 1, is composed of two classes defined by two manifolds, which are not linearly separable in this bi-dimensional space.

We have compared the proposed active learning method with random sampling and the recently proposed Bayesian method in [6]. Random sampling was implemented using the proposed method but, instead of selecting the samples according to Eq. (13), samples are selected randomly from the available training set. In all cases, a Gaussian kernel was used, whose optimal width parameter was selected by maximizing the standard cross-validation accuracy.

We divided the full set of 200 samples into two disjoint sets of 100 randomly selected samples each, one for training and the other for testing. We started our active learning process with a seed, a single labeled sample, randomly selected from the data set, that is, $M = 1$ at the beginning and the rest of the training set was used to simulate the oracle queries. We run the three algorithms for 99 iterations adding one sample at each iteration, that is, querying the oracle one sample each time so, at the end, $M = 100$. To obtain meaningful results, the process was repeated 10 times with different randomly selected training and test sets.

The performance of the algorithms is measured utilizing the samples in the test set using the mean confusion matrix, the mean overall accuracy (OA) and OA variance, and the mean kappa index. Each cell (i, j) of the mean confusion matrix contains the mean number of samples, over the ten executions of the algorithms using the different training and test sets, belonging to the j-th class, classified in the i-th class. The overall accuracy is the proportion of correctly

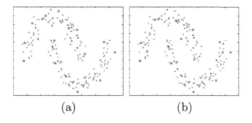

(a)   (b)

Fig. 1. First 15 selected samples for (a) the method in [6] and (b) the proposed method

classified samples over the total number of samples. The mean OA averages the ten OA results of the ten different algorithm executions. The variance of the OA in all the executions is reported as OA variance. The kappa index is a statistical measure, which reflects agreement between the obtained accuracy and the accuracy that would be expected by randomly classifying the samples. Unlike the Overall Accuracy, the kappa index avoids the chance effect. A value of the kappa index greater than 0.8 is considered to be "very good". Since ten runs of the algorithm are performed, the mean kappa over all the executions index is used.

In Figure 1 we show the first 15 selected samples for the method in [6] and the proposed method. It can be seen that both algorithms select samples that efficiently represent the two manifolds. Figure 2 shows the average learning curves for random sampling, the method in [6] and the proposed method. From the figure, it is clear that random sampling provides the lowest convergence rate, while the method in [6] and the proposed method have a similar learning rate to the full set overall accuracy. At convergence, when 100 samples are included in the training set, all methods have the same accuracy but the proposed method reaches this value with 18.4 samples on the average while the method in [6] needs 28.2 samples and random sampling needs 36.4 samples.

In the second experiment a real remote sensing dataset was used. Satellite or airborne mounted sensors usually capture a set of images of the same area in several wavelengths or spectral channels forming a multispectral image. This multispectral image allows for the classification of the pixels in the scene into different classes to obtain classification maps used for management, policy making and monitoring. A critical problem in remote sensing image segmentation is that few labeled pixels are typically available: in such cases, active learning may be very helpful [14].

We evaluated the methods on a real Landsat 5 TM image, whose RGB bands are depicted in Fig. 3a. The region of interest is a 1024×1024 pixels area centered in the city of Granada, in the south of Spain. The Landsat TM sensor provides a six bands multispectral image that covers RGB, near-infrared and mid-infrared ranges with a spatial resolution of 30 meters per pixel, that is, each pixels captures the energy reflected by the Earth in a square area of side equal to 30 meters. The dataset, created by the RSGIS Laboratory at the University of Granada, divides the scene into two classes, vegetation and no-vegetation. Note that the

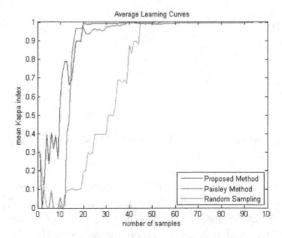

Fig. 2. Average learning curves for the active learning techniques using random sampling, the Bayesian method in [6] (Paisley method), and the proposed method for the synthetic experiment

(a)   (b)   (c)

Fig. 3. (a) Multispectral image, (b) classification map with the proposed method, and (c) classification map with the method in [6]. Pixels classified as vegetation are shown in green color and pixels classified as no-vegetation are shown in brown.

no-vegetation class includes bare soil that has a very similar spectral signature to vegetation making the correct classification of the pixels a challenging problem.

A total of 336 samples, whose class is precisely known by visual inspection of the images and by terrain inspection, were selected from the image, 174 samples corresponding to the vegetation class and 162 samples corresponding to the no-vegetation class. Each sample has six characteristics, each one corresponding to the mean value of a 3×3 area centered in the pixel under study for each one of the six bands that comprise the multispectral information provided by the Landsat TM satellite. Again, the same Gaussian kernel was used for all methods.

From the labeled dataset a test set of 150 samples was randomly selected, and the remaining 186 samples were used to simulate the oracle queries. We run the experiments 10 times with different training and test sets. All the algorithms

Table 1. Mean confusion matrix, mean kappa index, mean overall accuracy and its variance for ten runs of the method in [6] on different test sets

Predicted/actual	vegetation	no-vegetation	
vegetation	74.4	3.5	**Mean Kappa** = 0.9453
no-vegetation	0.6	71.5	**Mean OA** = 97.27%

Mean Kappa = 0.9453
Mean OA = 97.27%
OA variance = 4.39×10^{-5}

Table 2. Mean confusion matrix, mean kappa index, mean overall accuracy and its variance for ten runs of the proposed method on different test sets

Predicted/actual	vegetation	no-vegetation	
vegetation	74.4	2.4	**Mean Kappa** = 0.96
no-vegetation	0.6	72.6	**Mean OA** = 98.00%

Mean Kappa = 0.96
Mean OA = 98.00%
OA variance = 9.87×10^{-5}

were run for 185 iterations, starting from a training set with a single labeled pixel, that is $M = 1$, and adding one pixel to the training set at each iteration (query).

Again, the proposed method is compared with random sampling and the Bayesian method in [6]. For the method in [6] we did not perform the basis selection step. We want to note that, since this basis selection procedure discards features from the samples, better results are expected when all the features are used although the computational cost will be higher.

Figure 4 shows the average learning curves. The method in [6] provides a lower convergence rate to the full set overall accuracy than the proposed method. However, the method in [6] starts learning faster than the proposed one. It may be due to the fact that the active learning is carried out in an M-dimensional feature space while the proposed method works in an infinite-dimensional space. However, at convergence, when 186 samples have been included in the training set, the proposed method performs better than the method in [6]. Note also that, at convergence, random sampling obtains the same results with the proposed method, obtaining better classification accuracy than the method in [6]. This was expected since it uses the same classification procedure as the proposed method, except for the active learning selection procedure. Note, however, that the convergence rate is much slower than the other two methods.

Figures 3b and 3c depict the classification map for the full image using the proposed method and the method in [6]. The random sampling classification is not shown since, at convergence, coincides with the proposed method. The mean of the confusion matrices as well as the mean kappa index, the mean overall accuracy, and the overall accuracy variance are shown in Tables 1 and 2, for the method in [6] and the proposed method, respectively. From these figures of merit it is clear that the proposed method discriminates better between vegetation and no-vegetation than the method in [6].

All compared methods were implemented using Matlab© and run on a Intel i7 @ 2.67GHz. The proposed method took 1.23 sec to complete the 185 iterations while the method in [6] took 48.44 sec and random sampling took 1.01 sec. It

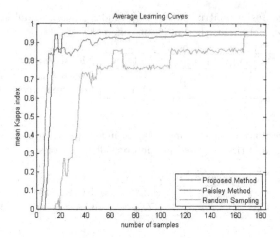

Fig. 4. Learning curve for the active learning techniques using random sampling, the Bayesian method in [6] (Paisley method), and the proposed method for the real remote sensing dataset

is worth noting that computing the precision matrix $\Sigma_{\mathbf{y},y(\mathbf{x}_+)|\gamma^2,\sigma^2}^{-1}$ in Eq. (16) takes most of the time, which explains the similar cost between the proposed method and random sampling. It is worth noting that the proposed method provided better figures of merit than the method in [6] for both mean kappa index and mean overall accuracy, learning with less interaction with the oracle and, also, with a much lower computational cost.

5   Conclusions

We presented an active learning procedure that exploits Bayesian learning and parameter estimation to tackle the problem of two-class kernel-based data classification. Using the Bayesian modeling and inference, we developed a Bayesian method for classification both finite and infinite dimensional feature spaces. The proposed method allows us to derive efficient closed-form expressions for parameter estimation and incremental and active learning. The method was experimentally compared to other methods and its performance was assessed on remote sensing multispectral image as well as synthetic data.

Acknowledgments. This work has been supported by the Spanish research programme Consolider Ingenio 2010: MIPRCV (CSD2007-00018) and the "Consejería de Innovación, Ciencia y Empresa of the Junta de Andalucía" under contract P07-TIC-02698. We want to thank V. F. Rodríguez-Galiano and Prof. M. Chica from the RSGIS laboratory (Group RNM122 of the Junta de Andalucía), who are supported by the Spanish MICINN (CGL2010-17629), for the image of the neighborhood of the city of Granada and the classified samples that conformed the real dataset used in this paper.

References

1. Babacan, D., Molina, R., Katsaggelos, A.: Bayesian compressive sensing using Laplace priors. IEEE Transactions on Image Processing 19(1), 53–63 (2010)
2. Bishop, C.M.: Pattern Recognition and Machine Learning (Information Science and Statistics). Springer (2007)
3. Burges, C.J.: A tutorial on support vector machines for pattern recognition. Data Mining and Knowledge Discovery 2, 121–167 (1998)
4. Elad, M.: Sparse and Redundant Representations - From Theory to Applications in Signal and Image Processing. Springer (2010)
5. MacKay, D.J.C.: Information-based objective functions for active data selection. Neural Computation 4(4), 590–604 (1992)
6. Paisley, J., Liao, X., Carin, L.: Active learning and basis selection for kernel-based linear models: A Bayesian perspective. IEEE Transactions on Signal Processing 58, 2686–2700 (2010)
7. Rasmussen, C.E., Williams, C.K.: Gaussian Processes for Machine Learning. MIT Press, NY (2006)
8. Schölkopf, B., Smola, A.: Learning with Kernels. MIT Press, Cambrigde (2002)
9. Seeger, M.W., Nickisch, H.: Compressed sensing and Bayesian experimental design. In: International Conference on Machine Learning 25 (2008)
10. Settles, B.: Active learning literature survey. Computer Sciences Technical Report 1648, University of Wisconsin–Madison (2009)
11. Shawe-Taylor, J., Cristianini, N.: Kernel Methods for Pattern Analysis. Cambridge Univ. Press (2004)
12. Suykens, J.A.K., Van Gestel, T., De Brabanter, J., De Moor, B., Vandewalle, J.: Least Squares Support Vector Machines. World Scientific, Singapore (2002)
13. Tipping, M.E.: The relevance vector machine. Journal of Machine Learning Research 1, 211–244 (2001)
14. Tuia, D., Volpi, M., Copa, L., Kanevski, M., Muñoz-Marí, J.: A survey of active learning algorithms for supervised remote sensing image classification. IEEE J. Sel. Topics Signal Proc. 4, 606–617 (2011)

Unsupervised Classification of SAR Images Using Hierarchical Agglomeration and EM

Koray Kayabol*, Vladimir A. Krylov**,
and Josiane Zerubia

Ariana, INRIA Sophia Antipolis Mediterranee,
2004 route des Lucioles, BP93, 06902 Sophia Antipolis Cedex, France
{koray.kayabol,vladimir.krylov,josiane.zerubia}@inria.fr
http://www-sop.inria.fr/ariana/en/index.php

Abstract. We implement an unsupervised classification algorithm for high resolution Synthetic Aperture Radar (SAR) images. The foundation of algorithm is based on Classification Expectation-Maximization (CEM). To get rid of two drawbacks of EM type algorithms, namely the initialization and the model order selection, we combine the CEM algorithm with the hierarchical agglomeration strategy and a model order selection criterion called Integrated Completed Likelihood (ICL). We exploit amplitude statistics in a Finite Mixture Model (FMM), and a Multinomial Logistic (MnL) latent class label model for a mixture density to obtain spatially smooth class segments. We test our algorithm on TerraSAR-X data.

Keywords: High resolution SAR, TerraSAR-X, classification, texture, multinomial logistic, Classification EM, hierarchical agglomeration.

1 Introduction

Finite Mixture Model (FMM) is a suitable statistical model to represent SAR image histogram and to perform a model based classification [1], [2]. One of the first uses of FMM in SAR image classification may be found in [3]. A combination of several different probability density functions (pdfs) into a FMM has been used in [4] for high resolution SAR images. The EM algorithm [5], [6] has been used for parameter estimation in latent variable models such as FMM. Two drawbacks of FMM based classification with EM algorithm can be sorted as 1) determination of the necessary number of class to represent the data and 2) initialization of the classes. By the term *unsupervised*, we refer to an initialization independent algorithm which is also able to determine the model order as in [7], [8]. There are some stochastic methods used in image segmentation like Reversible Jump Markov Chain Monte Carlo in [9], but their computational complexity is high

---

* Koray Kayabol carried out this work during the tenure of an ERCIM "Alain Bensoussan" Postdoctoral Fellowship Programme.
** Vladimir A. Krylov carried out this work with the support of INRIA Postdoctoral Fellowship.

E. Salerno, A.E. Çetin, and O. Salvetti (Eds.): MUSCLE 2011, LNCS 7252, pp. 54–65, 2012.

and sometimes they may reach an over segmented maps. Using the advantage of categorical random variables [10], we prefer to use an EM based algorithm called Classification EM (CEM) [11] whose computational cost is lower than both the stochastic methods and the conventional EM algorithm. In classification step, CEM uses the Winner-Take-All principle to allocate each data pixel to the related class according to the posterior probability of latent class label. After the classification step of CEM, we estimate the parameters of the class densities using only the pixels which belong to that class.

Running the EM type algorithms several times for different model orders to determine the order based on a criterion is a simple approach to reach a parsimonious solution. In [12], a combination of hierarchal agglomeration [13], EM and Bayesian Information Criterion (BIC) [14] is proposed to find the necessary number of classes in the mixture model. [8] performs a similar strategy with Component-wise EM [15] and Minimum Message Length (MML) criterion [16,17]. In this study, we combine hierarchal agglomeration, CEM and ICL [18] criterion to obtain an unsupervised classification algorithm.

We use a mixture of Nakagami densities for amplitude modelling. To obtain smooth and segmented class label maps, a post-processing can be applied to roughly classified class labels, but a Bayesian approach allows to include smoothing constraints to classification problems. We assume that each latent class label is a categorical random variable which is a special version of the multinomial random variable where each pixel belongs to only one class [1]. We introduce a spatial interaction within each binary map adopting multinomial logistic model [19], [10] to obtain a smooth segmentation map.

In Section 2 and 3, the MnL mixture model and CEM algorithm are given. We give the details of the agglomeration based unsupervised classification algorithm in Section 4. The simulation results are shown in Section 5. Section 6 presents the conclusion and future work.

2 Multinomial Logistic Mixture of Amplitude Based Densities

We assume that the observed amplitude at the nth pixel, $s_n \in \mathbb{R}^+$, where $n \in \mathcal{R} = \{1, 2, \ldots, N\}$ represents the lexicographically ordered pixel index, is free from any noise and instrumental degradation. Every pixel in the image has a latent class label. Denoting K the number of classes, we encode the class label as a K dimensional categorical vector $\mathbf{z}_n$ whose elements $z_{n,k}$, $k \in \{1, 2, \ldots, K\}$ have the following properties: 1) $z_{n,k} \in \{0, 1\}$ and 2) $\sum_{k=1}^{K} z_{n,k} = 1$. We may write the probability of s_n as the marginalization of the joint probability of $p(s_n, \mathbf{z}_n | \Theta, \boldsymbol{\pi}_n) = p(s_n | \mathbf{z}_n, \Theta) p(\mathbf{z}_n | \boldsymbol{\pi}_n)$, [1], as

$$p(s_n | \Theta, \boldsymbol{\pi}_n) = \sum_{\mathbf{z}_n} p(s_n | \mathbf{z}_n, \Theta) p(\mathbf{z}_n | \boldsymbol{\pi}_n)$$

$$= \sum_{\mathbf{z}_n} \prod_{k=1}^{K} [p(s_n | \theta_k) \pi_{n,k}]^{z_{n,k}} \tag{1}$$

where $\pi_{n,k} = p(z_{n,k} = 1)$ represent the mixture proportions, $\boldsymbol{\pi}_n = [\pi_{n,1}, \ldots, \pi_{n,K}]$, θ_k is the parameter of the class density and $\Theta = \{\theta_1, \ldots, \theta_K\}$ is the set of the parameters. If $\mathbf{z}_n$ is a categorical random vector and the mixture proportions are spatially invariant, (1) is reduced to classical FMM as follow:

$$p(s_n | \Theta) = \sum_{k=1}^{K} p(s_n | \theta_k) \pi_k \qquad (2)$$

We prefer to use the notation in (1) to show the contribution of the multinomial density of class label, $p(\mathbf{z}_n)$, into finite mixture model more explicitly. We give the details of the class and the mixture densities in the following two sections.

2.1   Class Amplitude Densities

Our aim is to use the amplitude statistics to classify the SAR images. For this purpose, we model the class amplitudes using Nakagami density, which is a basic theoretical multi-look amplitude model for SAR images [2]. We express the Nakagami density with parameters μ_k and ν_k as in [2], [10] as

$$p_A(s_n | \mu_k, \nu_k) = \frac{2}{\Gamma(\nu_k)} \left(\frac{\nu_k}{\mu_k} \right)^{\nu_k} s_n^{2\nu_k - 1} e^{\left(-\nu_k \frac{s_n^2}{\mu_k} \right)}. \qquad (3)$$

We denote $\theta_k = \{\mu_k, \nu_k\}$.

2.2   Mixture Density - Class Prior

The prior density $p(z_{n,k} | \pi_{n,k})$ of the categorical random variable is naturally an iid multinomial density, but we are not able to obtain a smooth class label map if we use an iid multinomial. We need to use a density which models the spatial smoothness of the class labels as well. We use a contrast function which emphasizes the high probabilities while attenuating the low ones, namely Logistic function [19]. The logistic function allows us to make an easier decision by discriminating the probabilities closed to each other. In this model, We are also able to introduce spatial interactions of the categorical random field by defining a binary spatial auto-regression model. Our MnL density for the problem at hand is written as

$$p(\mathbf{z}_n | \mathbf{Z}_{\partial n}, \eta) = \prod_{k=1}^{K} \left(\frac{\exp(\eta v_k(z_{n,k}))}{\sum_{j=1}^{K} \exp(\eta v_j(z_{n,j}))} \right)^{z_{n,k}} \qquad (4)$$

where

$$v_k(z_{n,k}) = 1 + \sum_{m \in \mathcal{M}(n)} z_{m,k}. \qquad (5)$$

and $\mathbf{Z}_{\partial n} = \{\mathbf{z}_m : m \in \mathcal{M}(n), m \neq n\}$ is the set which contains the neighbors of $\mathbf{z}_n$ in a window $\mathcal{M}(n)$ defined around n. The function $v_k(z_{n,k})$ returns the number of labels which belong to class k in the given window. The mixture density in (4) is spatially-varying with given function $v_k(z_{n,k})$ in (5).

3   Classification EM Algorithm

Since our purpose is to cluster the observed image pixels by maximizing the marginal likelihood given in (1), we suggest to use EM type algorithm to deal with the summation. The EM log-likelihood function is written as

$$Q_{EM}(\Theta|\Theta^{t-1}) = \sum_{n=1}^{N} \sum_{k=1}^{K} z_{n,k} \log\{p(s_n|\theta_k)\pi_{n,k}\}p(z_{n,k}|s_n, \mathbf{Z}_{\partial n}, \Theta^{t-1}) \quad (6)$$

where we include the parameter η to parameter set $\Theta = \{\theta_1, \ldots, \theta_K, \eta\}$.

If we used the exact EM algorithm to find the maximum of $Q(\Theta|\Theta^{t-1})$ with respect to Θ, we would need to maximize the parameters for each class given the expected value of the class labels. Instead of this, we use the advantage of working with categorical random variables and resort to Classification EM algorithm [11]. After classification step, we can partition the pixel domain $\mathcal{R}$ into K non-overlapped regions such that $\mathcal{R} = \bigcup_{k=1}^{K} \mathcal{R}_k$ and $\mathcal{R}_k \bigcap \mathcal{R}_l = 0, k \neq l$ and consequently, we may write the classification log-likelihood function as

$$Q_{CEM}(\Theta|\Theta^{t-1}) = \sum_{k=1}^{K} \sum_{m \in \mathcal{R}_k} \log\{p(s_m|\theta_k)\pi_{m,k}\}p(z_{m,k}|s_m, \mathbf{Z}_{\partial m}, \Theta^{t-1}) \quad (7)$$

The CEM algorithm incorporates a classification step between the E-step and the M-step which performs a simple Maximum-a-Posteriori (MAP) estimation to find the highest probability class label. Since the posterior of the class label $p(z_{n,k}|s_n, \mathbf{Z}_{\partial n}, \Theta^{t-1})$ is a discrete probability density function of a finite number of classes, we can perform the MAP estimation by choosing the maximum class probability. We summarize the CEM algorithm for our problem as follows:

E-step: For $k = 1, \ldots, K$ and $n = 1, \ldots, N$, calculate the posterior probabilities

$$p(z_{n,k}|s_n, \mathbf{Z}_{\partial n}, \Theta^{t-1}) = p(s_n|\theta_k^{t-1})\frac{\exp(\eta^{t-1}v_k(z_{n,k}))}{\sum_{j=1}^{K} \exp(\eta^{t-1}v_j(z_{n,j}))} \quad (8)$$

given the previously estimated parameter set Θ^{t-1}.

C-step: For $n = 1, \ldots, N$, classify the nth pixel into class j as $z_{n,j} = 1$ by choosing j which maximizes the posterior $p(z_{n,k}|s_n, \mathbf{Z}_{\partial n}, \Theta^{t-1})$ over $k = 1, \ldots, K$ as

$$j = \arg\max_k p(z_{n,k}|s_n, \mathbf{Z}_{\partial n}, \Theta^{t-1}) \quad (9)$$

M-step: To find a Bayesian estimate, maximize the classification log-likelihood in (7) with respect to Θ as

$$\Theta^{t-1} = \arg\max_\Theta Q_{CEM}(\Theta|\Theta^{t-1}) \quad (10)$$

To maximize this function, we alternate among the variables μ_k, v_k and η. We use the following methods to update the parameters: analytical solution of the first derivative for μ_k, zero finding of the first derivative for v_k and Newton-Raphson update equation for η.

4 Unsupervised Classification Algorithm

In this section, we present the details of the unsupervised classification algorithm. Our strategy follows the same general philosophy as the one proposed in [13] and developed for mixture model in [8,12]. We start the CEM algorithm with a large number of classes, $K = K_{max}$, and then we reduce the number of classes to $K \leftarrow K - 1$ by merging the weakest class in probability to the one that is most similar to it with respect to a distance measure. The weakest class may be found using the average probabilities of each class as

$$k_{weak} = \arg\min_k \frac{1}{N_k} \sum_{n \in R_k} p(z_{n,k}|s_n, \mathbf{Z}_{\partial n}, \Theta^{t-1}) \qquad (11)$$

Kullback-Leibler (KL) type divergence criterions are used in hierarchical texture segmentation for region merging [20]. We use a symmetric KL type distance measure called Jensen-Shannon divergence [21] which is defined between two probability density functions, i.e. $p_{k_{weak}}$ and p_k, $k \neq k_{weak}$, as

$$D_{JS}(k) = \frac{1}{2} D_{KL}(p_{k_{weak}}||q) + \frac{1}{2} D_{KL}(p_k||q) \qquad (12)$$

where $q = 0.5 p_{k_{weak}} + 0.5 p_k$ and

$$D_{KL}(p||q) = \sum_k p(k) \log \frac{p(k)}{q(k)} \qquad (13)$$

We find the closest class to k_{weak} as

$$l = \arg\min_k D_{JS}(k) \qquad (14)$$

and merge them to constitute a new class $R_l \leftarrow R_l \bigcup R_{k_{weak}}$.

We repeat this procedure until we reach the predefined minimum number of classes K_{min}. We determine the necessary number of classes by observing the ICL criterion explained in Section 4.3. The details of the initialization and the stopping criterion of the algorithm are presented in Section 4.1 and 4.2. The summary of the algorithm can be found in Table 1.

4.1 Initialization

The algorithm can be initialized by determining the class areas manually in case that there are a few number of classes. We suggest to use an initialization strategy for completely unsupervised classification. It removes the user intervention from the algorithm and enables to use the algorithm in case of large number of classes. First, we run the CEM algorithm for one global class. Using the cumulative distribution of the fitted Nakagami density $g = F_A(s_n|\mu_0, \nu_0)$ where $g \in [0, 1]$ and dividing $[0, 1]$ into K equal bins, we can find our initial class parameters as $\mu_k = F_A^{-1}(g_k|\mu_0, \nu_0)$, $k = 1, \ldots, K$ where g_k's are the centers of the bins. We initialize the other parameters using the estimated parameters of the global class.

Table 1. Unsupervised CEM algorithm for classification of amplitude based mixture model.

Initialize the classes defined in Section 4.1 for $K = K_{max}$.
While $K \geq K_{min}$, do

    While the condition in (15) is false, do
        E-step: Calculate the posteriors in (8)
        C-step: Classify the pixels regarding to (9)
        M-step: Estimate the parameters of amplitude and
        texture densities as (10)
    Find the weakest class using (11)
    Find the most similar class to the weakest class using
    (12-14)
    Merge these two classes $\mathcal{R}_l \leftarrow \mathcal{R}_l \bigcup \mathcal{R}_{k_{weak}}$
    $K \leftarrow K - 1$

4.2 Stopping Criterion

We observe the normalized and weighted absolute difference between sequential values of parameter set θ_k to decide the convergence of the algorithm. We assume that the algorithm has converged, if the following expression is satisfied:

$$\sum_{k=1}^{K} \frac{N_k |\theta_k^t - \theta_k^{t-1}|}{N |\theta_k^{t-1}|} \leq 10^{-3} \tag{15}$$

4.3 Choosing the Number of Classes

Although the SAR images which we used have a small number of classes, we validate our assumption on number of classes using the Integrated Completed Likelihood (ICL) [18]. Even though BIC is the most used and the most practical criterion for large data sets, we prefer to use ICL because it is developed specifically for classification likelihood problem, [18], and we have obtained better results than BIC in the determination of the number of classes. In our problem, the ICL criterion may be written as

$$ICL(K) = \sum_{n=1}^{N} \sum_{k=1}^{K} \log\{p(s_n|\hat{\theta}_k)^{\hat{z}_{n,k}} p(\hat{z}_{n,k}|\hat{\mathbf{Z}}_{\partial n}, \hat{\eta})\} - \frac{1}{2} d_K \log N \tag{16}$$

where d_K is the number of free parameters. In our case, it is $d_K = 12 * K + 1$. $\hat{z}_{n,k}$ is the maximum a posterior estimate of $z_{n,k}$ found in C-step.

We also use the BIC criterion for comparison. It can be written as

$$BIC(K) = \sum_{n=1}^{N} \log \left(\sum_{k=1}^{K} p(s_n|\hat{\theta}_k) p(z_{n,k}|\mathbf{Z}_{\partial n}, \hat{\eta}) \right) - \frac{1}{2} d_K \log N \tag{17}$$

5   Simulation Results

This section presents the high resolution SAR image classification results of the proposed method called AML-CEM (Amplitude density mixtures of MnL with CEM), compared to the corresponding results obtained with other methods which are DSEM-MRF [22] and K-MnL. We have also tested supervised version of AML-CEM [10] where training and testing sets are determined by selecting some spatially disjoint class regions in the image, and we run the algorithm twice for training and testing. The sizes of the windows for texture and label models are selected to be 3×3 and 13×13 respectively by trial and error. We initialize the algorithm as described in Section 4.1 and estimate all the parameters along the iterations.

The K-MnL method is the sequential combination of K-means clustering for classification and Multinomial Logistic label model for segmentation to obtain a fairer comparison with the K-means clustering since K-means does not provide any segmented map. The weak point of the K-means algorithm is that it does not converge to the same solution every time, since it starts with random seed. Therefore, we run the K-MnL method 20 times and select the best result among them.

We tested the algorithms on the following TerraSAR-X image:

- TSX1: 1200 × 1000 pixels, HH polarized TerraSAR-X Stripmap (6.5 m ground resolution) 2.66-look geocorrected image which is acquired over San-chagang, China (see Fig. 1(a)). ©Infoterra.

For TSX1 image in Fig.1(a), the full ground-truth map has been generated manually. Fig.1 shows the classification results where the red colored regions indicate the misclassified parts according to 3-classes ground-truth map. We can see the plotted ICL and BIC values with respect to number of classes in Fig. 2. The variations in the ICL and BIC plots are slowed down after 3. ICL reaches its maximum value at 4. Since the difference between the values at 3 and 4 is very small and our aim is to find the minimum number of classes, we may say that the mixture model with 3 number of classes is almost enough to represent this data set. Fig. 3 shows several classification maps found with different numbers of classes. From this figure, we can see the evolution of the class maps along the agglomeration based algorithm. The numerical accuracy results are given in Table 2 for 3-classes. The supervised AML-CEM gives the best result over all. In supervised methods, the accuracy results of K-MnL is a bit better than those of unsupervised AML-CEM, but K-MnL is able to provide these results in case of a given number of classes. Due to this, we may call K-MnL as a semi-supervised method.

The simulations were performed on MATLAB platform on a PC with Intel Xeon, Core 8, 2.40 GHz CPU. The total 57 iterations are performed in 5.07 minutes for $K = 1, \ldots, 8$ to plot the graphic in Fig. 2 and the total number of the processed pixels are about 1.2 millions.

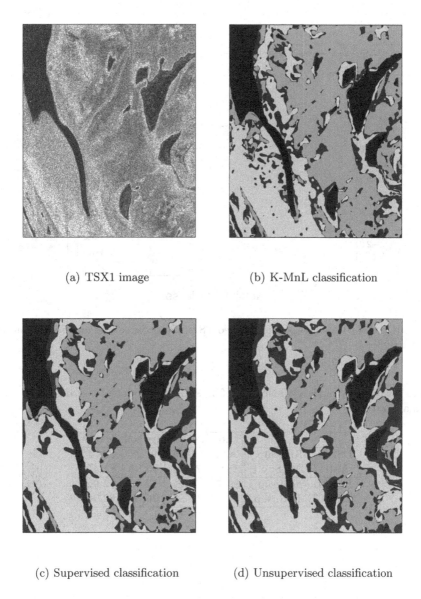

(a) TSX1 image   (b) K-MnL classification

(c) Supervised classification   (d) Unsupervised classification

Fig. 1. (a) TSX1 image, (b), (c) and (d) classification maps obtained by K-MnL, supervised and unsupervised ATML-CEM methods. Dark blue, light blue, yellow and red colors represent water, wet soil, dry soil and misclassified areas, respectively.

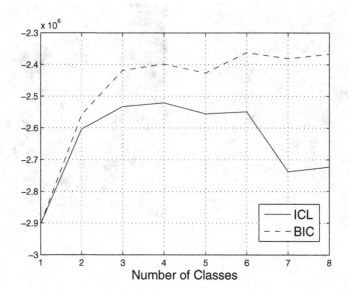

Fig. 2. ICL and BIC values of the classified TSX1 image for several numbers of sources (from 1 to 8)

Table 2. Accuracy of the classification of TSX1 image in water, wet soil and dry soil areas and average

	water	wet soil	dry soil	average
DSEM-MRF (Sup.)	**90.00**	69.93	91.28	83.74
AML-CEM (Sup.)	88.98	**71.21**	**93.06**	**84.42**
K-MnL (Semi-sup.)	**89.71**	**86.13**	72.42	**82.92**
AML-CEM (Unsup.)	88.24	62.99	**96.39**	82.54

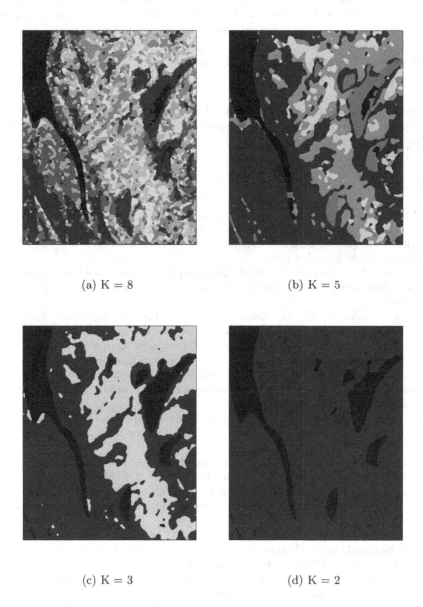

(a) K = 8   (b) K = 5

(c) K = 3   (d) K = 2

Fig. 3. Classification maps of TSX1 image obtained with unsupervised ATML-CEM method for different numbers of classes K = {2,3,5,8}

6   Conclusion and Future Work

Using an agglomerative type unsupervised classification method, we eliminate
the negative effect of the latent class label initialization. According to our experiments, the larger number of classes, we start the algorithm with, the more initial
value independent results, we obtain. Consequently, the computational cost is
increased as a by-product. The classification performance may be increased by
including additional features as texture or polarization.

Acknowledgments. The authors would like to thank Aurélie Voisin (Ariana
INRIA, France) for interesting discussions and Astrium-Infoterra GmbH for providing the TerraSAR-X image.

References

1. Titterington, D., Smith, A., Makov, A.: Statistical Analysis of Finite Mixture Distributions, 3rd edn. John Wiley & Sons, Chichester (1992)
2. Oliver, C., Quegan, S.: Understanding Synthetic Aperture Radar Images, 3rd edn.
Artech House, Norwood (1998)
3. Masson, P., Pieczynski, W.: SEM Algorithm and Unsupervised Statistical Segmentation of Satellite Images. IEEE Trans. Geosci. Remote Sens. 31(3), 618–633
(1993)
4. Krylov, V.A., Moser, G., Serpico, S.B., Zerubia, J.: Supervised Enhanced
Dictionary-Based SAR Amplitude Distribution Estimation and Its Validation With
Very High-Resolution Data. IEEE Geosci. Remote Sens. Lett. 8(1), 148–152 (2011)
5. Dempster, A.P., Laird, N.M., Rubin, D.B.: Maximum Likelihood from Incomplete
Data via the EM Algorithm. J. R. Statist. Soc. B. 39, 1–22 (1977)
6. Redner, R.A., Walker, H.F.: Mixture Densities, Maximum Likelihood and the EM
Algorithm. SIAM Review 26(2), 195–239 (1984)
7. Palubinskas, G., Descombes, X., Kruggel, F.: An Unsupervised Clustering Method
using the Entropy Minimization. In: Int. Conf. Pattern Recognition, ICPR 1998,
pp. 1816–1818 (1998)
8. Figueiredo, M.A.T., Jain, A.K.: Unsupervised Learning of Finite Mixture Models.
IEEE Trans. on Pattern Anal. Machine Intell. 24(3), 381–396 (2002)
9. Wilson, S.P., Zerubia, J.: Segmentation of Textured Satellite and Aerial Images
by Bayesian Inference and Markov Random Fields. Res. Rep. RR-4336, INRIA,
France (2001)
10. Kayabol, K., Voisin, A., Zerubia, J.: SAR Image Classification with Non-stationary
Multinomial Logistic Mixture of Amplitude and Texture Densities. In: Int. Conf.
Image Process, ICIP 2011, pp. 173–176 (2011)
11. Celeux, G., Govaert, G.: A Classification EM Algorithm for Clustering and Two
Stochastic Versions. Comput. Statist. Data Anal. 14, 315–332 (1992)
12. Fraley, C., Raftery, A.: Model-based Clustering, Discriminant Analysis, and Density Estimation. J. Am. Statistical Assoc. 97(458), 611–631 (2002)
13. Ward, J.H.: Hierarchical groupings to optimize an objective function. J. Am. Statistical Assoc. 58(301), 236–244 (1963)
14. Schwarz, G.: Estimating the Dimension of a Model. Annals of Statistics 6, 461–464
(1978)

15. Celeux, G., Chretien, S., Forbes, F., Mkhadri, A.: A Component-wise EM Algorithm for Mixtures. Res. Rep. RR-3746, INRIA, France (1999)
16. Wallace, C.S., Boulton, D.M.: An Information Measure for Classification. Comp. J. 11, 185–194 (1968)
17. Wallace, C.S., Freeman, P.R.: Estimation and Inference by Compact Coding. J. R. Statist. Soc. B 49(3), 240–265 (1987)
18. Biernacki, C., Celeux, G., Govaert, G.: Assessing a Mixture Model for Clustering with the Integrated Completed Likelihood. IEEE Trans. on Pattern Anal. Machine Intell. 22(7), 719–725 (2000)
19. Krishnapuram, B., Carin, L., Figueiredo, M.A.T., Hartemink, A.J.: Sparse Multinomial Logistic Regression: Fast Algorithms and Generalization Bounds. IEEE Trans. on Pattern Anal. Machine Intell. 27(6), 957–968 (2005)
20. Scarpa, G., Gaetano, R., Haindl, M., Zerubia, J.: Hierarchical Multiple Markov Chain Model for Unsupervised Texture Segmentation. IEEE Trans. Image Process. 18(8), 1830–1843 (2009)
21. Lin, J.: Divergence Measures Based on the Shannon Entropy. IEEE Trans. Inform. Theory 37(1), 145–151 (1991)
22. Krylov, V.A., Moser, G., Serpico, S.B., Zerubia, J.: Supervised High-Resolution Dual-Polarization SAR Image Classification by Finite Mixtures and Copulas. IEEE J. Sel. Top. Signal Process. 5(3), 554–566 (2011)

Geometrical and Textural Component Separation with Adaptive Scale Selection

Tamás Szirányi[1] and Dániel Szolgay[2]

[1] Computer and Automation Research Institute, MTA SZTAKI, Budapest, Hungary
sziranyi@sztaki.hu
[2] Pázmány Péter Catholic University, Budapest, Hungary
szoda@digitus.itk.ppke.hu

Abstract. The present paper addresses the cartoon/texture decomposition task, offering theoretically clear solutions for the main issues of adaptivity, structure enhancement and the quality criterion of the goal function. We apply Anisotropic Diffusion with a Total Variation based adaptive parameter estimation and automatic stopping condition. Our quality measure is based on an observation that the cartoon and the texture components of an image are orthogonal to each other. The visual and numerical comparison to the similar algorithms from the state-of-the-art showed the superiority of the proposed method.

1 Introduction

The definition of the details and that of the main structure of images is a challenging problem in several image analysis tasks. The Scale-space theory[1, 6] gives us fundamental answers for this issue, but the discrimination of fine details and main outlines remained a further important issue. Denoising [9, 11] , defocus [5] and cartoon/texture filters are three main areas where important details enhanced against the remaining image tissues. The present paper addresses the cartoon/texture discrimination tasks, offering theoretically clear solutions for the main issues of adaptivity, structure enhancement and the quality criterion of the goal function.

Decomposition of images into texture and non-texture (or cartoon) components can be useful for image compression [13], denoising [9, 11], feature selection[18], etc. Most of the recently published algorithms in the field [3, 12, 14, 18, 19] are based on total variation (TV) minimization inspired by the work of Yves Meyer [7]. Total variation based regularization dates back to Tikhonov [16]. The most widely known form was introduced in image processing by Mumford and Shah[8] for image segmentation and later by Rudin et al. [11] for noise removal through the optimization of a cost function as follows:

$$\inf \left\{ E_{TV}(u) = \int_{\Omega} |Du| + \lambda \int_{\Omega} v^2 \right\} \qquad (1)$$

where u is the cartoon component of the original image f, $v = f - u$ is the texture, $\int_{\Omega} |Du|$ denotes the total variation of u in Ω and λ is a regularization

E. Salerno, A.E. Çetin, and O. Salvetti (Eds.): MUSCLE 2011, LNCS 7252, pp. 66–77, 2012.
© Springer-Verlag Berlin Heidelberg 2012

parameter. The first part produces a smooth image with bounded variation upon energy minimization for the cartoon component, while the second ensures that the result is close to the initial image. The regularization of Rudin et al. [11] (**ROF** in the following) was used as an image denoising and deblurring method, since it removes fine, oscillating, noise-like patterns, but preserves sharp edges.

In [7] Meyer proposed a different norm for the second, texture part of (1), which is better suited for oscillatory components than the standard L_2 norm:

$$\inf \int_\Omega |Du| + \lambda \|v\|_* \tag{2}$$

where $\|.\|_*$ is defined on a suitable Banach G space as follows:

$$\|v\|_* = \inf_{g1,g2} \left\| \sqrt{g_1^2(x,y) + g_2^2(x,y)} \right\|_{L^\infty} \tag{3}$$

over all $g_1, g_2 \in L^\infty(\mathbb{R}^2)$ such that $v = div(g)$ where $g = (g_1, g_2)$. Other variations of eq.(1) are summarized in [3].

Beside the choice of regularization, other techniques are used to enhance the quality of the decomposition: in [19], the authors propose an image decomposition and texture segmentation method via sparse representation using Principal Component Analysis (**DPCA**). In [12], an algorithm (**DOSV**) is introduced to find the optimal value of the fidelity parameter (λ in eq.(1)) based on the observation of Aujol et al. in [2] concerning the independence of cartoon and texture.

Looking at the palette of the different solutions, we can see that the decomposition into cartoon and textured partitions requires tackling the following main issues:

- Adaptive scale definition of texture and cartoon (cc. outline) details;
- Process that filters out textured parts while keeping the main outlines;
- Quality criterion for the efficiency of the decomposition: a goal function.

In the following, we overview the related contributions and then we introduce our proposed solutions to the above issues.

1.1   Related Works

In this section, we shortly summarize published results closely related to the proposed method: non-linear filtering is introduced in [3], Anisotropic Diffusion in [10, 17] and measures of independence in [2, 5, 15].

BLMV Nonlinear Filter. Buades et al. have recently proposed a non-linear method (based on the name of the authors it will be called **BLMV** filter in the following) that calculates local total variation (LTV) for each pixel on f before and after filtering the image with a σ sized low pass filter, L_σ, inspired by Y. Meyer[7]. The relative difference of the calculated LTVs shows if the observed pixel is part of the texture or the cartoon. The cartoon image, u is composed based on this information: if the relative difference is high for a pixel r, then $u(r)$ will be equal to the low pass filtered $(L_\sigma * f)(r)$, otherwise $u(r) = f(r)$. The results of this simple method are impressive on the presented examples in [3]: the edges are preserved as long as σ is not too large and the texture components are blurred with L_σ.

The right choice of σ is important to get the best result, however, usually there is no such σ which eliminates all the textures but keeps the non-texture components on the cartoon. The existence of a content adaptive scaling parameter can be derived from scale-space theory, as it has been introduced in Lindeberg's works [6].

Anisotropic Diffusion. The general goal of diffusion algorithms is to remove noise from an image by using partial differential equations (PDE). Diffusion algorithms can be classified as isotropic or anisotropic. Isotropic diffusion is equivalent to using a Gaussian filter on the image, which blurs not only the noise or texture components, but the main edges as well.

In [10] Perona and Malik proposed an Anisotropic Diffusion (**AD**) functions that, according to scale-space theory (see works of Alvarez, Lions and Morel [1]) allows diffusion along the edges or in edge-free territories, but penalizes diffusion orthogonal to the edge direction:

$$\frac{\partial f(x,y,t)}{\partial t} = div(g(\|\nabla f\|)\nabla f) \tag{4}$$

where $f(x,y,t) : \mathbb{R}^2 \to \mathbb{R}^+$ is the image in the continuous domain, with (x,y) spatial coordinates, t is an artificial time parameter ($f(x,y,0)$ is the original image). ∇f is the image gradient, $\|\nabla f\|$ is the magnitude of the gradient and $g(.)$ is the weighting function that controls diffusion along and across edges. The discretized form of their diffusion equation is as follows:

$$f(x,y,t+1) = f(x,y,t)+$$
$$+ \frac{\lambda}{|\eta(x,y)|} \sum_{(x',y')\in\eta(x,y)} \nabla^{(x',y')} g(\|\nabla^{(x',y')}f(x,y,t)\|)\nabla^{(x',y')}f(x,y,t) \tag{5}$$

where f is the processed image, (x,y) is a pixel position, t now denotes discrete time steps (iterations). The constant $\lambda \in \mathbb{R}^+$ is a scalar that determines the rate of diffusion, $\eta(x,y)$ is the spatial neighborhood of (x,y), $|\eta(x,y)|$ is the number of neighboring pixels. $\nabla^{(x',y')}f(x,y,t)$ is an approximation of the image gradient at a particular direction:

$$\nabla^{(x',y')}f(x,y,t) = f(x',y',t) - f(x,y,t), (x',y') \in \eta(x,y) \tag{6}$$

AD belongs to a theoretically sound scale-space class of differential processes ensuring the denoising of an image along with the enhancement of its main structure [4]. We will show that the AD as proposed by Perona and Malik is not suitable for cartoon/texture decomposition, since the texture part might contain high magnitude edges, which would inhibit the diffusion. As a solution to this problem, the authors of [13] suggest to use the AD algorithm with modified weights: instead of using $\|\nabla^{(x',y')}f(x,y,t)\|$ as the parameter of the weighting function, they use the edges of the Gaussian filtered image, $\nabla(G_\sigma * f)$:

$$\|\nabla^{(x',y')}(G_\sigma * f)(x',y',0) - (G_\sigma * f)(x,y,0)\|, (x',y') \in \eta(x,y) \qquad (7)$$

where G_σ is a Gaussian filter with σ bandwidth and $*$ denotes the convolution. Using a blurred image to control diffusion directions will give better results: texture edges will not hinder the diffusion, but the strong main edges will do. Yet the quality of the solution relies heavily on the σ parameter: with small σ, some texture might remain on the cartoon, while with greater σ, some of the cartoon edges will disappear.

In Section 2 we will propose an algorithm which utilizes the smoothing property of AD, while it helps preserving edges based on whether they belong to a cartoon or texture and not based on their level of magnitude.

The Use of Independence in Image Decomposition. The independence of the carton part and the texture/noise part of the image was used in denoising, decomposition[2] and restoration [15] algorithms.

In [2], Aujol et al. propose the use of the correlation between the cartoon and the oscillatory (noise, texture) components of a decomposition to estimate the regularization parameter λ. The assumption made in their model is that these two components are uncorrelated, which makes intuitive sense (as stated in [12]), since every feature of an image should be considered as either a cartoon feature or a textural/noise feature, but not both.

In [15], the Angle Deviation Error (ADE) - introduced in [5] – is used as a measure of independence to automatically find the best stopping point for an iterative non-regularized image deconvolution method. The described ADE measure is somewhat similar to correlation [2], but it is based on the pure orthogonality of two image partitions (e.g. clear image and noise):

$$ADE(Q,P) = \left|\arcsin\left(\frac{\langle Q,P\rangle}{|Q| \cdot |P|}\right)\right| \qquad (8)$$

where $Q, P \in \mathbb{R}^n$ and $\langle Q,P\rangle$ is their scalar product. This measure is different from the standard correlation, where zero-mean vectors are used to calculate the scalar product and the normalization is done with the standard deviation of the vectors:

$$corr(Q,P) = \frac{cov(Q,P)}{\sigma_Q \cdot \sigma_P} = \frac{\sum_{i=1}^{n}(Q_i - \mu_Q)(P_i - \mu_P)}{n \cdot \sigma_Q \cdot \sigma_P} \qquad (9)$$

where $cov(.)$ is the covariance over the elements of vectors, σ_Q, μ_Q and σ_P, μ_P are the standard deviation and the expected values of the elements of Q and P respectively, and n is the size of the vectors.

Comparing the two measures, we can see that they are very similar: if both Q and P were zero mean, the two measures would actually give the same result.

However, in cartoon texture decomposition only the texture part has an inherent zero mean, while the cartoon does not. This makes a small difference in the resulting decomposed images in favor of the ADE measure, as it will be shown in Section 3; ADE strengthens the image partitions to being really independent (geometrical orthogonality in $\mathbb{R}^n$), while $corr$ is for the estimation of regression.

1.2 The Contribution of the Paper

In the following we will show how independence can be used to better separate the texture and cartoon parts of the image by using ADE orthogonality measure to locally estimate the best parameter of the BLMV filter. The edge inhibitions of the AD are initialized by the filtered image. Then the ADE is calculated again on the diffused image to stop the diffusion at the point where the orthogonality of cartoon and texture components is maximal. To sum up, we offer theoretically clear solutions for the main issues:

- Adaptive scale definition by using locally optimal BLMV filter tuned by ADE measure;
- Anisotropic Diffusion, initialized by the new adaptive BLMV to better separate texture from cartoon;
- Orthogonality criterion for the quality measure of the decomposition (stopping condition to AD).

In the following we overview in detail the proposed solutions for the above issues. To validate the our method, we will show results on real life images, and also on artificial images where numerical evaluation is possible.

2 Cartoon/Texture Decomposition Using Independence Measure

In this section, the orthogonality based cartoon/texture decomposition method is described in detail. The core algorithm is the AD, which is initialized and stopped using BLMV filter and ADE independence measure.

2.1 Locally Adaptive BLMV Filter

As it has been mentioned earlier, the BLMV filter uses the same σ sized low pass filter for the whole image, while there is no guarantee for the existence of a single sigma that would remove all texture from the image without blurring the cartoon edges.

We propose the use of different σ for the different parts of the image based on the independence of the removed texture component and the remaining cartoon component. This theory is similar to the one proposed in [2], although in our case the parameter selection has to be locally adaptive. The reason for this difference lies in the purpose of the methods: while in [2] the goal of the authors was noise removal, where one can assume that the parameters of the noise are the same for the whole image, here we want to remove texture components which may vary in many aspects (e.g. scale, magnitude) across the image.

To make the filter locally adaptive, BLMV filtered images were calculated for a given range of the scale parameter: $\sigma_i \in [s_1; s_2]$. Let $u_{\sigma_i}, v_{\sigma_i}$ denote the cartoon and texture components of the f input image, produced by the BLMV filter with σ_i parameter.

The image is then divided into non-overlapping small cells (5 pixel by 5 pixel in our experiments), and around each cell a larger block (21 by 21) is centered, in which the ADE measure is calculated:

$$ADE(u_{\sigma_i}^{(x,y)}, v_{\sigma_i}^{(x,y)}) = \left| \arcsin \left(\frac{\langle u_{\sigma_i}^b(x,y) v_{\sigma_i}^b(x,y) \rangle}{|u_{\sigma_i}^b(x,y)| \cdot |v_{\sigma_i}^b(x,y)|} \right) \right|, \tag{10}$$

where $u^b(x, y)$ and $v^b(x, y)$ denote the cartoon and texture components of the *block* which is centered around the cell containing (x, y) pixel.

It is worth noting that the texture component of an image should have zero mean, since it is the difference of the textured area and the diffused background. To eliminate the consequences of the quantization error through the iterations, the texture component is biassed to have zero mean when the ADE function is computed.

The σ with minimal ADE is chosen to be the parameter for each pixel in the cell. For the output cartoon image the value of the pixel, $u_a(x, y)$ will be the following:

$$u_a(x, y) = u_{\sigma_m^{(x,y)}}(x, y) \tag{11}$$

$$\sigma_m^{(x,y)} = \underset{\sigma_i \in [s_1; s_2]}{\arg \min} \left(ADE \left(u_{\sigma_i}^b(x, y), v_{\sigma_i}^b(x, y) \right) \right) \tag{12}$$

This cell-based scheme is used to reduce the computational workload: instead of calculating the block correlation for each pixel, we calculate it for small cells. To avoid the blocking effect, a soft Gaussian smoothing was used on the parameter image of the same size as the input image and containing the corresponding σ value in each (x, y) point: $p(x, y) = \sigma_m^{(x,y)}$.

2.2 Anisotropic Diffusion with Adaptive BLMV Filter and ADE Stopping Condition

The above described adaptive BLMV filter (aBLMV in the following) clearly performs better than the original one (see Section 3), but it still faces a problem at the borders where cartoon and texture parts meet: either the cartoon edges are blurred, or the texture remains on the cartoon component close to cartoon edges. We propose to use AD initialized with a cartoon image produced by aBLMV filter and stopped by ADE measure. AD preserves high magnitude edges and blurs weaker ones, but obviously a texture can contain strong edges while a cartoon edge can be weak. As a result, AD may blur important edges of the cartoon and keep unwanted edges of the texture.

Similarly to [13], where the diffusion weight function was calculated on a Gaussian blurred version of the image, we propose to calculate the $g(.)$ weight function of eq.(5) by using the aBLMV-filtered image resulting in the following diffusion equation:

$$f(x, y, t + 1) = f(x, y, t)+$$

$$+ \frac{\lambda}{|\eta(x,y)|} \sum_{(x',y') \in \eta(x,y)} \nabla^{(x',y')} g(\nabla^{(x',y')} u_a(x, y)) \nabla^{(x',y')} f(x, y, t) \qquad (13)$$

On u_a of eq.(11), the texture parts are blurred and they do not contain strong edges, while the cartoon parts are more or less preserved. Choosing a low value for the rate of diffusion (λ) means that the diffusion can preserve even the weak edges of the cartoon part, but it blurs texture parts completely (since it is not inhibited by edges).

To avoid oversmoothing of important edges, the iteration of the AD must be stopped at the right moment. For this purpose, we utilize the independence property of cartoon and texture components in the same manner as we did in Section 2.1, with the difference that here we are searching for the iteration count i that minimizes $ADE(u_i, v_i)$ for each block.

The cartoon component of the proposed method is produced as follows:

$$u(x, y) = f(x, y, t_{ADE}), \qquad (14)$$

$$t_{ADE} = \arg \min_{i=1..I_{max}} (ADE(f^b(x, y, i), v^b(x, y, i))) \qquad (15)$$

where $f(x, y, t_{ADE})$ is the (x, y) pixel of the diffused image after t_{ADE} iterations, I_{max} is the maximum number of diffusion iterations, $f^b(x, y, i)$ and $v^b(x, y, i) = f^b(x, y, 0) - f^b(x, y, i)$ are the cartoon and texture components, respectively, of the *block* that is centered around the cell containing (x, y) pixel after the ith diffusion iteration.

If the diffusion is not stopped automatically, but after fixed number of iterations, then some parts of the cartoon component will be apparent on the texture image.

3   Results

The evaluation of the quality of cartoon/texture decomposition is usually done on visual examples, since there is no generally accepted objective method for ground truth generation in case of real images. Sometimes it is difficult even for a human to decide if a certain part of the image is texture or not. Hence, to evaluate the quality of the different methods, we show the decomposition results of example images (see Fig.1), but we also evaluate numerically the competing methods on artificial images (see Fig.2) where the ground truth cartoon and texture parts are available.

(a) Barbara (787x576, ROI: 301x301)   (b) Dublin (300x300)

Fig. 1. Images used for visual evaluation. In parentheses after the image name there is the size of the image and the size of the Region of Interest (ROI) if the latter examples are using only a part of the image.

We have compared the proposed method to the following decomposition methods: BLMV-filter [3], aBLMV-filter (also proposed in this paper), DPCA [19], DOSV [12], TVL1 [18]. The codes for the above methods were provided by the authors, and we used them with the best tuned parameters in each individual test case. For numerical evaluation, we used the parameters that gave the best figures, and in case of subjective evaluation, the parameters that gave the best visual result. For the proposed method, we kept all the parameters unchanged: the maximum number of iterations for the AD was set to 100, the λ scalar parameter of eq. (5) was set to 2. The $[s_1, s_2]$ range of the σ was set to $s_1 = 0pix$ and $s_2 = 7pix$.

Fig. 2. The artificial images used for numerical evaluation. Top row: original image, bottom row: cartoon component.

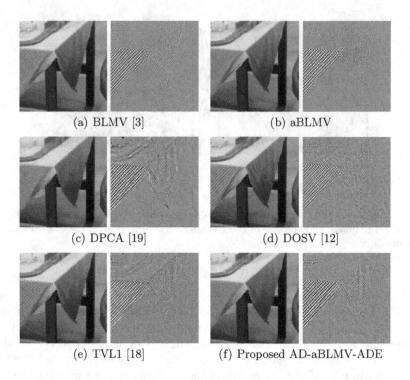

(a) BLMV [3]   (b) aBLMV

(c) DPCA [19]   (d) DOSV [12]

(e) TVL1 [18]   (f) Proposed AD-aBLMV-ADE

Fig. 3. Separation of cartoon and texture components (Barbara)

For better visibility, the contrast of the texture images was linearly stretched on the demonstrated figures.

For the visual evaluation, one has to consider how strong the remaining cartoon parts on the texture image and the remaining texture part on the cartoon image are. For a part of the Barbara image, we can see on Fig.3 that 3 methods (BLMV, TVL1, DOSV) cannot completely eliminate the texture from the table cover, while there are cartoon edges apparent on the texture image. DPCA can eliminate the texture from the cartoon image, but the image itself becomes less smooth, and the slow changes of gray level values are also apparent on the texture image. The BLMV with adaptive local parameter selection (aBLMV) and the proposed method eliminate the texture from the cartoon while virtually no cartoon appear on the texture image (see Fig.3).

The second image, the Dublin shows similar results. See Fig.4. BLMV, aBLMV cannot eliminate the texture parts completely without blurring the cartoon. TVL1 and DPCA perform similarly: they both eliminate most of the texture, but a lot of non-textural edges are apparent on the texture image. DOSV and the proposed method keeps the texture part the most "clear", but DOSV leaves some part of the texture on the cartoon image.

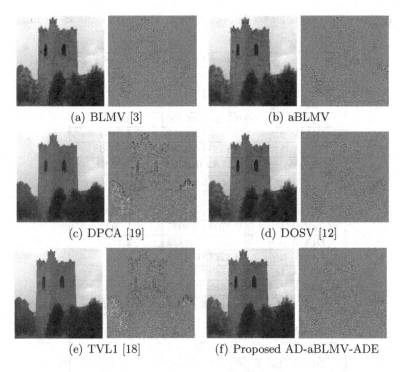

(a) BLMV [3]   (b) aBLMV

(c) DPCA [19]   (d) DOSV [12]

(e) TVL1 [18]   (f) Proposed AD-aBLMV-ADE

Fig. 4. Separation of cartoon and texture components (Dublin)

Numerical evaluation is a difficult task for cartoon/texture decomposition since usually there is no ground truth for the images. For this reason most papers in the field lack this kind of comparison and rely only on subjective visual evaluation. We used artificial images for numerical evaluation where the ground truth is available. The following measures were calculated to compare quality: edge absolute difference of the cartoon ($ead(u)$) and texture ($ead(v)$) images, and the absolute difference of the cartoon image ($ad(u)$). We define these measures as follows:

$$ead(u) = |e(u') - e(u)|, \quad ead(v) = |e(v') - e(v)|, \quad ad(u) = |u' - u|, \quad (16)$$

where u and v are the ground truth cartoon and texture images, u', v' are the cartoon and texture images produced by a decomposition method and $e(.)$ is the Prewitt edge image. In general the proposed AD-aBLMV-ADE method performs better than the rest (see TABLE 1).

We have also compared the proposed method using ADE to the case when correlation is used instead as an independence measure. It shows that the pure orthogonality criterion performs slightly better than the correlation based comparison (see TABLE 2).

Table 1. Numerical results for the images of Fig.2. The best results are highlighted in bold

	Scores for the 1st image of Fig.2		
	ead(u)	ead(v)	ad(u)
BLMV [3]	0.8393	0.7611	2.6356
aBLMV	0.7922	0.7182	2.4525
DPCA [19]	0.7420	0.6596	2.8667
DOSV [12]	0.4137	0.3911	1.5184
TVL1 [18]	0.4339	0.3288	3.8850
AD-aBLMV-ADE	**0.2782**	**0.2408**	**1.0342**
	Scores for the 2nd image of Fig.2		
	ead(u)	ead(v)	ad(u)
BLMV [3]	0.4363	0.3229	1.8498
aBLMV	0.3703	0.3037	1.5932
DPCA [19]	0.5535	0.4322	2.3272
DOSV [12]	0.3217	0.2788	1.6535
TVL1 [18]	0.5589	0.4391	2.2239
AD-aBLMV-ADE	**0.2682**	**0.2385**	**1.2826**
	Scores for the 3rd image of Fig.2		
	ead(u)	ead(v)	ad(u)
BLMV [3]	0.4479	0.3147	1.4710
aBLMV	0.3979	0.2805	1.4539
DPCA [19]	0.6842	0.5437	9.2985
DOSV [12]	0.3564	0.3090	1.2509
TVL1 [18]	0.6917	0.4942	29.2620
AD-aBLMV-ADE	**0.1610**	**0.1304**	**1.2245**

Table 2. Ratio of the error rates and the correlation of ADE based vs. Correlation based calculus. The results obtained by ADE are better: all three absolute differences have decreased.

ADE vs. Corr		
ead(u)	ead(v)	ad(u)
97.34%	97.31%	98.73%

4   Conclusion

In this paper, we have addressed the main issues of the cartoon/texture decomposition task, offering theoretically clear solutions for the adaptivity, structure enhancement and the quality criterion of the goal function, using anisotropic diffusion with ADE based iteration stopping. To initialize the diffusion inhibitions of AD, we used BLMV [3] nonlinear filter with adaptive parameter selection based on ADE calculus. Numerical results and visual comparisons show that the proposed method works with high efficiency and in quality it outperforms other algorithms introduced in recent years.

Acknowledgment. We would like to thank the authors of [3, 12, 18, 19] for making their algorithms available for testing.

This work was partially supported by the Hungarian Research Fund #76159.

References

1. Alvarez, L., Lions, P.L., Morel, J.M.: Image selective smoothing and edge detection by nonlinear diffusion. ii. SIAM J. Numer. Anal. 29, 845–866 (1992)
2. Aujol, J.F., Gilboa, G.: Constrained and snr-based solutions for tv-hilbert space image denoising. J. Math. Imaging Vis. 26, 217–237 (2006)
3. Buades, A., Le, T., Morel, J.M., Vese, L.: Fast cartoon + texture image filters. IEEE Tr. on Image Processing 19(8), 1978–1986 (2010)
4. Kopilovic, I., Sziranyi, T.: Artifact reduction with diffusion preprocessing for image compression. Optical Engineering 44(2) (2005)
5. Kovacs, L., Sziranyi, T.: Focus area extraction by blind deconvolution for defining regions of interest. IEEE Tr. Pattern Analysis and Machine Intelligence 29(6), 1080–1085 (2007)
6. Lindeberg, T.: Edge detection and ridge detection with automatic scale selection. Int. Journal of Computer Vision 30, 117–154 (1998)
7. Meyer, Y.: Oscillating Patterns in Image Processing and Nonlinear Evolution Equations: The Fifteenth Dean Jacqueline B. Lewis Memorial Lectures. American Mathematical Society, Boston (2001)
8. Mumford, D., Shah, J.: Optimal approximations by piecewise smooth functions and associated variational problems. Communications on Pure and Applied Mathematics 42(5), 577–685 (1989)
9. Osher, S., Sole, A., Vese, L.: Image decomposition and restoration using total variation minimization and the h^{-1} norm. Sci. Multiscale Model. Simul. 1, 349–370 (2002)
10. Perona, P., Malik, J.: Scale-space and edge detection using anisotropic diffusion. IEEE PAMI 12, 629–639 (1990)
11. Rudin, L.I., Osher, S., Fatemi, E.: Nonlinear total variation based noise removal algorithms. Phys. D 60, 259–268 (1992)
12. Shahidi, R., Moloney, C.: Decorrelating the structure and texture components of a variational decomposition model. IEEE Tr. on Image Processing 18(2), 299–309 (2009)
13. Sprljan, N., Mrak, M., Izquierdo, E.: Image compression using a cartoon-texture decomposition technique. In: Proc. WIAMIS, p. 91 (2004)
14. Starck, J.L., Elad, M., Donoho, D.: Image decomposition via the combination of sparse representations and a variational approach. IEEE Tr. on Image Processing 14(10), 1570–1582 (2005)
15. Szolgay, D., Sziranyi, T.: Optimal stopping condition for iterative image deconvolution by new orthogonality criterion. Electronics Letters 47(7), 442–444 (2011)
16. Tikhonov, A.N., Arsenin, V.Y.: Solutions of ill-posed problems. Scripta series in mathematics, Winston, Washington (1977)
17. Weickert, J.: Anisotropic Diffusion in Image Processing. Teubner-Verlag, Stuttgart (1998)
18. Yin, W., Goldfarb, D., Osher, S.: Image Cartoon-Texture Decomposition and Feature Selection Using the Total Variation Regularized L1 Functional. In: Paragios, N., Faugeras, O., Chan, T., Schnörr, C. (eds.) VLSM 2005. LNCS, vol. 3752, pp. 73–84. Springer, Heidelberg (2005)
19. Zhang, F., Ye, X., Liu, W.: Image decomposition and texture segmentation via sparse representation. IEEE Sig. Processing Letters 15, 641–644 (2008)

Bayesian Shape from Silhouettes

Donghoon Kim and Rozenn Dahyot

School of Computer Science and Statistics
Trinity College Dublin, Ireland
{Donghook,Rozenn.Dahyot@tcd.ie}
http://www.scss.tcd.ie

Abstract. This paper extends the likelihood kernel density estimate of the visual hull proposed by Kim et al [1] by introducing a prior. Inference of the shape is performed using a meanshift algorithm over a posterior kernel density function that is refined iteratively using both a multiresolution framework (to avoid local maxima) and using KNN for selecting the best reconstruction basis at each iteration. This approach allows us to recover concave areas of the shape that are usually lost when estimating the visual hull.

Keywords: shape from silhouettes, kernel density estimates, Mean-shift algorithm, Gaussian stack, KNN, visual hull.

1 Introduction

Three dimensional reconstruction of an object that is seen by multiple image sensors, has many applications such as 3d modelling [2, 3] or video surveillance [4]. Shape from silhouettes methods are very popular in computer vision because of their simplicity and their computational efficiency. Laurentini [5] has defined the Visual Hull as the best reconstruction that can be computed using an infinite number of silhouettes captured from all viewpoints outside the convex hull of the object. Volume-based approaches focus on the volume of the visual hull [6, 7, 8, 5] while surface-based approaches, less numerically stable, aim at estimating a surface representation of the visual hull [9, 10, 11, 12].

Volume based approaches based on voxel occupancy rely on the optimisation of a discrete objective function. As an alternative, Kim et al. [1] have recently introduced a smooth objective function that can be optimised with Meanshift algorithm to infer 3D shapes from silhouettes. Their modelling can be understood as a likelihood function approximating the visual hull. The visual hull however is the convex envelope of the object and concave regions of the object are not reconstructed. We propose here to extend Kim et al.'s modelling by adding prior information (section 3) that allows us to recover these concave regions. We assess our method experimentally in section 4 for reconstructing a 2D shape from 1D binary silhouettes and we show that it is reconstructing the 2D shapes accurately even when few camera views are available. We also show an illustration for 3D reconstruction from 2D silhouettes using our posterior modelling before concluding in section 5.

E. Salerno, A.E. Çetin, and O. Salvetti (Eds.): MUSCLE 2011, LNCS 7252, pp. 78–89, 2012.

2   Context and Motivations

Volume-based approaches focus on the volume of the *Visual Hull* [6, 7, 8, 5]
The world volume is splitted in elementary blocks (voxel) and each block can
project onto a pixel in the recorded silhouettes. Like the bin of an histogram,
the block is incremented each time it projects on to the foreground part of a
silhouette image. Such representation corresponds to a histogram representation
as an approximation of the probability density function of the spatial random
variable $\mathbf{x}$ to be in the volume of the object. Such discrete approximation, even
when having collected an infinite number of silhouettes from all possible angle
of views, can only reconstruct the convex envelope of the object. The quality of
this reconstruction depends on the number of camera views, their viewpoints, the
voxel resolution and the complexity of the object. The calibration parameters of
each camera are needed to compute back-projection functions which allow the
positions in 3D space to be mapped to the image planes.

The discrete nature of the histogram makes the approach memory demanding. Moreover, optimisation methods (e.g. exhaustive search) of such discrete
representation are limited and suboptimal compared to smooth modellings that
can be optimised with gradient ascent methods. To aleviate this limitation, Kim
et al. [1] have recently proposed a smooth Kernel density estimate (KDE) as
another approximation of the probability density function of the spatial random
variable $\mathbf{x}$ to be in the volume of the object. For simplicity, they considered
that a 3D object volume is composed of a stack of 2D slices, and their KDE is
modelled for a spatial random variable $\mathbf{x}$ in a slice (i.e. $\mathbf{x} \in \mathbb{R}^2$). The estimation (search for maxima) is performed in each slice, and the volume is extracted
by combining the results for all slices. Their modelling was also limited to silhouettes recorded by orthographic cameras. Ruttle et al. [13] have extended this
modelling to using standard pin-hole cameras, and their inference is directly performed in the 3D world. Newton Raphson and Meanshift algorithms [1, 13] can
be used efficiently to search for the maxima of these KDEs. Indeed parallel implementations of these algorithms have been proposed on a Graphics Processing
Unit (GPU) [14, 15].

These smooth KDEs [1, 13] can be interpreted as likelihoods since they link
the latent variable (i.e. the spatial position of the object $\mathbf{x}$) and the observations
(silhouettes and camera parameters). However, without prior information about
the object to be reconstructed, these modellings give an estimate of the visual
hull and are therefore unable to reconstruct concave parts of the object. This
paper proposes to extend Kim et al. [1] approach by introducing a prior in the
modelling (section 3). The reconstruction is then performed by optimisation of
posterior by a gradient ascent technique (Meanshift). For simplicity, we use the
same hypotheses as Kim et al. [1]: the cameras are orthographic, their parameters
are known, and the modelling is done in 2D slices. Consequently our observations
correspond to 1D binary silhouettes, and the random variable $\mathbf{x} \in \mathbb{R}^2$ is in a 2D
slice. The likelihood is written as [1]:

$$lik(\mathbf{x}) \propto \sum_{i=1}^{n} \exp\left(\frac{-(\rho_i - \mathbf{x}^T\mathbf{n}_i)^2}{2h^2}\right) \pi_i \qquad (1)$$

where for pixel i in the silhouettes, $\pi_i = 0$ if it is on the background, and $\pi_i = 1$ if it is on the foreground. n is the total number of pixels seen in all silhouette images, and $(\rho_i, \mathbf{n}_i)$ corresponds to the orthographic camera parameter for pixel i.

3   Modelling of the Prior

Morphable models are commonly used to model a prior for a class of shapes [16]. As a alternative here, we model a new prior using K-neareast neighbours (sections 3.1 and 3.2). The resulting posterior, when the likelihood is combined with the prior, is updated in a coarse to fine fashion during the estimation of the shape (section 3.3).

3.1   Shape Description

The shape is described by a sequence of connected points. The points are chosen uniformly along the contour in an anti-clockwise direction. Note the sequence of points is normalised by subtracting the mean of the points coordinates, and by dividing by their respective variance. This is a standard pre-processing step for getting a representation invariant to translation and scale. Our shape descriptor contains not only the ordered list of 2D points $\{\mathbf{x}_i = (x_i, y_i)\}_{i=1,\cdots,M}$, but also its local angles $\{\alpha_i\}_{i=1,\cdots,M}$: α_i is the angle between the vectors $\mathbf{x}_i - \mathbf{x}_{i+1}$ and $\mathbf{x}_i - \mathbf{x}_{i-1}$. We define the function f between :

$$f(\mathbf{X}) = \begin{bmatrix} \alpha_1 \\ \alpha_2 \\ \alpha_3 \\ \vdots \\ \alpha_M \end{bmatrix} \text{ with } \mathbf{X} = \begin{bmatrix} \mathbf{x}_1 \\ \mathbf{x}_2 \\ \mathbf{x}_3 \\ \vdots \\ \mathbf{x}_M \end{bmatrix} \qquad (2)$$

where M is the number of sampled points to describe the shape. Note that $\mathbf{X}$ and $f(\mathbf{X})$ are not invariant to rotation i.e. choosing a different starting point $\mathbf{x}_1 = (x_1, y_1)$ on the shape will lead to other vectors $\mathbf{X}'$ and $f(\mathbf{X}')$ that will be cyclic permutation of $\mathbf{X}$ and $f(\mathbf{X})$. The representation $f(\mathbf{X})$ is however invariant to scale changes on $\mathbf{X}$.

3.2   Shape Prior Modelling with KNN

Having a new shape $\mathbf{X}$ and a training database of N shape exemplars $\{\mathbf{X}_j^e\}_j = 1, \ldots, N$, a standard modelling is to assume a linear relationship of the form:

$$\mathbf{X} = \sum_{k=1}^{K} \omega_k \mathbf{U}_k + \epsilon \qquad (3)$$

where the vectors $\{\mathbf{U}_k\}_{k=1,\cdots,K}$ are a selected basis of vectors computed from the training database and ϵ is the error between the observation and the reconstruction on this basis. The vectors $\{\mathbf{U}_k\}_{k=1,\cdots,K}$ are usually defined offline using for instance PCA. As an alternative, we choose to use the K-nearest neighbours instead where the nearest K exemplars are used to define a prior for $\mathbf{X}$. The distance metric between shapes $\mathbf{X}$ and $\mathbf{Y}$, is defined as follows:

$$d(\mathbf{X}, \mathbf{Y}) = \sum_{i=0}^{M} |\alpha_i^\mathbf{X} - \alpha_i^\mathbf{Y}| \tag{4}$$

This metric is an absolute distance between $f(\mathbf{X})$ and $f(\mathbf{Y})$ and is used to find the nearest neighbours. We define our basis of functions $\{\mathbf{U}_k\}_{k=1,\cdots,K}$ by selecting the K exemplars of the training database that will be at the shortest distance of a shape $\mathbf{X}$. To be insensitive to rotation, we also consider all cyclic permutation of the exemplars. The weights $\{\omega_k\}_{k=1,\cdots,K}$ (equation 3) are calculated as follows:

$$\omega_k = \frac{1}{(K-1)} \left(1 - \frac{d_k}{d_{sum}}\right) \tag{5}$$

where $d_{sum} = \sum_{k=1}^{K} d_k$ and $d_k = d(\mathbf{X}, \mathbf{U}_k)$. Note that by definition, the weights sum to 1, $\sum_{k=1}^{K} \omega_k = 1$. The reconstruction $\mathbf{X}_U$ is defined on the basis of the nearest neighbours:

$$\mathbf{X}_U = \sum_{k=1}^{K} \omega_k \, \mathbf{U}_k \tag{6}$$

Note however that the reconstruction $\mathbf{X}_U$ approximates the observed shape $\mathbf{X}$, but at a normalised scale since the exemplars in the training database are all normalised.

3.3   Posterior and Inference

Having a current guess of the shape noted $\hat{\mathbf{X}}^{(t)}$, we can compute the reconstruction $\mathbf{X}_U^{(t)} = [\mathbf{x}_{U_1}^{(t)}, \cdots, \mathbf{x}_{U_M}^{(t)}]$ using the K nearest neighbours in the training database of exemplars $\mathcal{S}_{prior}$. We model a prior using $\mathbf{X}_U^{(t)}$ to allow for the estimation of the refined shape at $(t+1)$. Each of the M points of $\mathbf{X}^{(t+1)} = [\mathbf{x}_1^{(t+1)}, \cdots, \mathbf{x}_M^{(t+1)}]$ is updated individually.

Lets consider the first point $\mathbf{x}_1^{(t+1)}$. The likelihood is modelled using the kernel density estimate (1) and, the prior for $\mathbf{x}_1^{(t+1)}$ is modelled given $\mathbf{X}_U^{(t)}$ and $\hat{\mathbf{X}}^{(t)}$:

$$post(\mathbf{x}_1^{(t+1)}) \propto lik(\mathbf{x}_1^{(t+1)}) \times prior(\mathbf{x}_1^{(t+1)} | \mathbf{X}_U^{(t)}, \hat{\mathbf{X}}^{(t)}) \tag{7}$$

The reconstruction $\mathbf{X}_U^{(t)}$ is converted into $M-1$ angles noted $\Theta_1^{(t)} = [\theta_2^{(t)}, \cdots, \theta_M^{(t)}]$ such that $\theta_m^{(t)}$ corresponds to the slope of the line defined by $(\mathbf{x}_{U_1}^{(t)}, \mathbf{x}_{U_m}^{(t)})$. $\theta_m^{(t)}$ is the orientation of the unitary vector $\mathbf{n}_m^{(t)} = (\cos(\theta_m^{(t)}), \sin(\theta_m^{(t)}))^T$. We assume that the update $\mathbf{x}_1^{(t+1)}$ is in the neighborhood of the line defined by the angle $\theta_m(t)$ going through the point $\hat{\mathbf{x}}_m^{(t)}$. This can be translated into the following equation:

$$\mathbf{n}_m^{(t)T}(\mathbf{x}_1^{(t+1)} - \hat{\mathbf{x}}_m^{(t)}) = \epsilon_p \qquad (8)$$

where $\epsilon_p \sim \mathcal{N}(0, h_p^2)$ is the error with Normal distribution (mean 0, variance h_p^2). In a similar fashion as for the likelihood, the prior is then modelled using a kernel density estimate:

$$prior(\mathbf{x}_1^{(t+1)}|\mathbf{X}_U^{(t)}, \hat{\mathbf{X}}^{(t)}) \propto \sum_{m=2}^{M} \exp\left(-\frac{(\mathbf{n}_m^T(\mathbf{x}_1^{(t+1)} - \hat{\mathbf{x}}_m^{(t)}))^2}{2h_p^2}\right), \qquad (9)$$

We use only slopes from the reconstruction and therefore this method is invariant to scale difference between the shape $\hat{\mathbf{X}}^{(t)}$ and the normalised reconstruction $\mathbf{X}_U^{(t)}$.

Since both the likelihood and the prior are kernel density estimates, the posterior distribution is also a KDE and the mean shift algorithm is used to maximise the posterior:

$$\hat{\mathbf{x}}_1^{(t+1)} = \arg\max_{\mathbf{x}_1^{(t+1)}} \left\{ post(\mathbf{x}_1^{(t+1)}) \right\} \qquad (10)$$

This is repeated for each point in the contour such that the estimated update is computed:

$$\hat{\mathbf{X}}^{(t+1)} = [\hat{\mathbf{x}}_1^{(t+1)}, \cdots, \hat{\mathbf{x}}_M^{(t+1)}]$$

The shape of the initial point set $\hat{\mathbf{X}}^{(0)}$ is the result of the estimation using only the likelihood [1]. In order to converge iteratively toward a good solution even so the starting guess $\hat{\mathbf{X}}^{(0)}$ is far from it, we need to be careful in the approximation of the prior. In particular, the selection of the K nearest neighbours may not be the best basis of functions for the reconstruction at the start since $\hat{\mathbf{X}}^{(0)}$ is only a convex approximation to the shape. To avoid this problem, we construct a Gaussian shape stack whose concept is introduced in Lefebvre and Hoppe [17]. The Gaussian stack is constructed by smoothing the exemplar shapes in the prior set using increasing bandwidths (noted $h_e^{(t)}$) without downsampling the shapes as it is usually done in Gaussian pyramids. This stack is computed using the convolution with a gaussian (with bandwidth $h_e^{(t)}$) on all exemplars in the training database from large to small bandwidth as a smoothing factor. We note $S_{prior}^{h_e(t)}$ the set of exemplars smoothed with a Gaussian of bandwidth $h_e^{(t)}$. The bandwidth $h_e^{(t)}$ decreases at each iteration of the mean shift algorithm as follows:

$$h_e^{(t)} = \alpha^t \ h_{max} \text{ until } h_e = h_{min} \text{ with } \alpha = 0.9 \qquad (11)$$

where, $h_{max} = 13$, $h_{min} = 1$, h_{max} is selected experimentally to be as similar as possible to the optimal initial point set. This procedure allows to achieve a coarse-to-fine strategy in modelling the prior. Figure 1 shows how an exemplar shape evolves from a convex smooth shape to a more structured one as the bandwidth

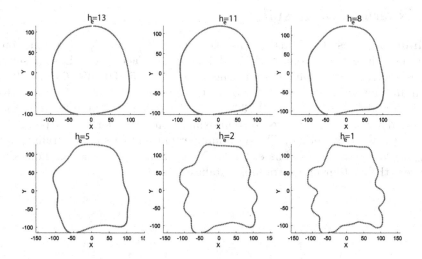

Fig. 1. Multiresolution approach: variations of one of the exemplars in the training database w.r.t. h_e

h_e decreases. The reconstruction at time t, $\hat{\mathbf{X}}^{(t)}$, that is approximated from the selected exemplars in the training database, is then iteratively refined to get more accurate shape estimation results. The estimation procedure is summarised in Algorithm 1.

Algorithm 1. Reconstruction from binary silhouettes using posterior

---

Compute of an initial guess $\hat{\mathbf{X}}^{(0)}$ of the shape at time $t = 0$ with the likelihood [1]
Init $h_e(0) = h_{max} = 13$
repeat
    Select the K nearest exemplars of $\hat{\mathbf{X}}^{(t)}$ in $\mathcal{S}_{prior}^{h_e(t)}$ and compute $\mathbf{X}_U^{(t)}$
    for $i = 1 \rightarrow M$ **do**
        Model the prior for $\mathbf{x}_i^{(t+1)}$ (eq. (9))
        $\hat{\mathbf{x}}_i^{(t+1)} = \arg\max_{\mathbf{x}_i^{(t+1)}} \left\{ post(\mathbf{x}_i^{(t+1)}) \propto lik(\mathbf{x}_i^{(t+1)}) \times prior(\mathbf{x}_i^{(t+1)} | \mathbf{X}_U^{(t)}, \hat{\mathbf{X}}^{(t)}) \right\}$
    end for
    $t \leftarrow t + 1$
    $h_e(t) = \alpha^t h_{max}$ with $\alpha = 0.9$
until $\hat{\mathbf{X}}^{(t+1)} \simeq \hat{\mathbf{X}}^{(t)}$ and $h_e(t) < 1$

Remarks. The proposed prior is updated iteratively so that concavity information can be introduced progressively by using a Gaussian stack, and it is also refined at each step by choosing the nearest neighbours of the current estimate (KNN approach). This strategy differs from standard approaches with PCA where the reconstruction is computed as a linear combination of K fixed pre-selected components. Our KNN-based approach on the contrary refines the selection of these components iteratively in our estimation process.

4   Experimental Results

Training and Test Databases. The 2D shapes to model the prior are the contours of 6 objects taken from the ALOI database [18]. Each object class has seven images recorded from different viewing angles $[0°, 15°, 30°, 45°, 60°, 75°, 90°]$ which are divided into the test database (three images $[15°, 45°, 75°]$) and the training database (four images $[0°, 30°, 60°, 90°]$) as can be seen in figure 2. The training database $\mathcal{S}_{prior}$ is used to approximate the prior for the shape in the k-nearest neighbors method. The total number of exemplars in the training set $\mathcal{S}_{prior}$ is $N = 6 \times 4 = 24$. The exemplar $\mathbf{X}^e$ is sampled in $M = 360$ points to represent the contour in the training database.

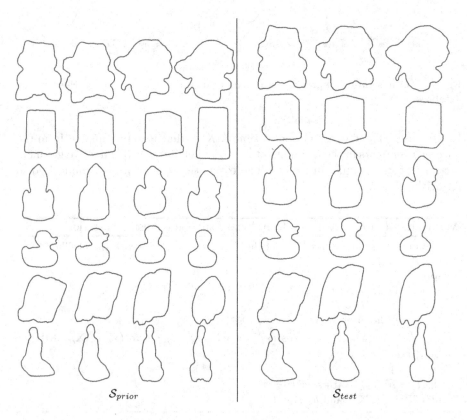

$$\mathcal{S}_{prior} \qquad\qquad\qquad \mathcal{S}_{test}$$

Fig. 2. Training database (left) $\mathcal{S}_{prior}$ (all cyclic permutations of these exemplars are also taken into account to allow the reconstruction process to be insensitive to rotations) and test database $\mathcal{S}_{test}$ (right)

Observations. The observed silhouettes correspond to 1D binary signals: the contours in $\mathcal{S}_{test}$ are back-projected using orthographic projection in different

directions. These projections are computed using the Radon transform that are
then thresholded to give binary silhouettes.

Experiments. We compare here the following methods:

- the likelihood using only silhouettes [1],
- the posterior modelled with the likelihood using silhouettes [1] and a shape
  prior approximated by the KNN method ($K = 2$).

The Euclidean distance is used to measure the distance between the ground-truth
and the estimated shapes. Figure 3 shows the mean of the euclidian distance and
standard error computed using the objects in S_{test} with respect to the number
of projections (i.e. camera views) available. The distance for the likelihood can
only decrease w.r.t. the number of silhouettes available up to a point where the
convex visual hull is recovered.

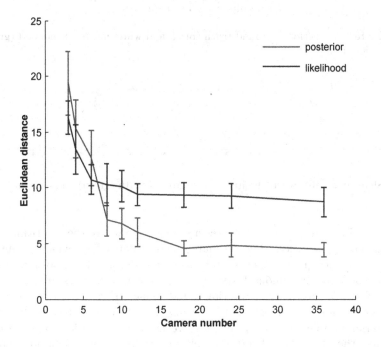

Fig. 3. Euclidean distance plot with standard error w.r.t. the number of camera views:
Kim et al. likelihood (blue) and the result using the posterior (red)

The reconstructed contours are shown in Figures 4 on top of the ground
truth. The proposed prior allows to recover concave regions reducing the distance
between the ground truth and the estimated shape. The likelihood only converges
toward the visual hull which is a convex envelope of the object. The top left result

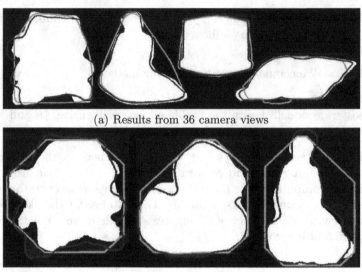

(a) Results from 36 camera views

(b) Results from 3 camera views

Fig. 4. 2D Reconstructions: Ground-truth (black and white image), likelihood (green) and posterior (blue)

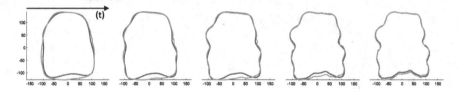

Fig. 5. Shape estimates w.r.t. the iteration t: ground truth (green) and estimates $\hat{\mathbf{X}}^{(t)}$ (red)

of the posterior (figure 4(a)) illustrates a special case where the algorithm leads to a non optimal solution. The estimated shapes depend on which exemplars are selected for designing the prior and this case is difficult for choosing optimal exemplars because the silhouettes from different viewing angles are similar for both the ground truth and the solution found. Hence there are not enough clues (in the observations) to choose the best exemplars. Overall the results show that the posterior is able to recover the concave parts of the object, and perform better than the likelihood when the number of camera view is superior to 5.

Figure 5 shows the estimates of the shape at different iterations t of our algorithm. Concavity is introduced iteratively by decreasing the smoothing parameter h_e, and we can see that the estimate $\hat{\mathbf{X}}^{(t)}$ is very close to the ground truth (i.e. corresponding smoothed exemplar).

Remarks. The proposed prior has been designed using standard ideas for reconstructing a contour on a basis of selected components. KNN was chosen here to

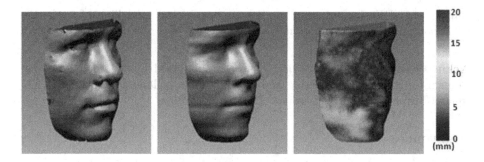

Fig. 6. Example of 3D reconstruction combining 2D slice: ground truth (left), reconstruction using the posterior (middle), distance between the estimate and the ground truth in mm (right)

select the basis of components but PCA could also have been used [19]. KNN was the most efficient method in this example because any shape in the test set will be best explained by the $K = 2$ neighbours from the prior set that correspond to the same object as viewed from a slightly different angle. For instance the duck viewed at angle $15°$ in the test set is very close to the two ducks viewed at angles $0°$ and $30°$ in the training set (cf. figure 2). Note that the proposed algorithm can then be adapted to any other strategy for finding the best components.

Extension to 3D. The inference scheme proposed in this paper can be used to reconstruct a 3D object from 2D silhouettes by combining the estimates on each individual 2D slices [19]. Figure 6 illustrates the result of using our posterior, using 36 binary silhouettes of an unknown face as inputs, and a prior designed on 35 faces generated using the Basel Face Model [20]. 70 slices along the vertical axis are used to represent the full shape. Note that discontinuous circular bands appear on the surface of the reconstruction due to the 2D slice representation (Fig. 6). However, simple post-processing like vertical smoothing can easily remove these discontinuities.

5   Conclusion and Future Work

This paper has proposed a Kernel density estimate of a posterior density function to infer shape from silhouettes. Optimisation is performed using the Mean-shift algorithm. The likelihood currently used is designed for silhouettes recorded with orthographic cameras but can easily be replaced to handle standard pin-hole cameras [13]. Starting from an initial guess of the shape, the prior is modelled using its K-nearest neighbours in the training exemplars. The shape estimate is refined iteratively (along with the prior). Experimental results shows that while the likelihood is limited to recover the convex hull of the shape, the posterior allows to recover its concave areas when enough non-ambiguous information is available from the observations. Future efforts will investigate a smooth modelling with KDEs including colour information in the modelling of both the likelihood and the prior (photohull).

Acknowledgments. This project has been supported by the Irish Research Council for Science, Engineering and Technology in collaboration with INTEL Ireland Ltd: funded by the National Development Plan.

References

[1] Kim, D., Ruttle, J., Dahyot, R.: 3d shape estimation from silhouettes using mean-shift. In: IEEE International Conference on Acoustics Speech and Signal Processing (ICASSP), pp. 1430–1433 (2010)

[2] Dyer, C.R.: Volumetric scene reconstruction from multiple views. In: Davis, L.S. (ed.) Foundation of Image Understanding, pp. 469–489. Kluwer, Boston (2001)

[3] Franco, J.S., Boyer, E.: Efficient polyhedral modeling from silhouettes. IEEE Transactions on Pattern Analysis and Machine Intelligence, 414–427 (2008)

[4] Zimmermann, K., Svoboda, T., Matas, J.: Multiview 3d tracking with an incrementally constructed 3d model. In: The Third International Symposium on 3D Data Processing, Visualization, and Transmission (2006)

[5] Laurentini, A.: How far 3d shapes can be understood from 2d silhouettes. IEEE Transactions on Pattern Analysis and Machine Intelligence 17(2), 188–195 (1995)

[6] Martin, W.N., Aggarwal, J.K.: Volumetric description of objects from multiple views. IEEE Transactions on Pattern Analysis and Machine Intelligence 5(2), 150–158 (1987)

[7] Potmesil, M.: Generating octree models of 3d objects from their silhouettes in a sequence of images. Computer Vision, Graphics, and Image Processing 40, 1–29 (1987)

[8] Srivasan, P., Liang, P., Hackwood, S.: Computational geometric methods in volumetric intersections for 3d reconstruction. Pattern Recognition 23(8), 843–857 (1990)

[9] Baumgart, B.G.: Geometric modeling for computer vision. Technical report, Artificial Intelligence Laboratory, Stanford University (1974)

[10] Lazebnik, S., Boyer, E., Ponce, J.: On computing exact visual hulls of solids bounded by smooth surfaces. In: IEEE Conference on Computer Vision and Pattern Recognition, vol. 1, pp. 156–161 (2001)

[11] Matusik, W., Buehler, C., McMillan, L.: Polyhedral visual hulls for real-time rendering. In: Eurographics Workshop on Rendering, pp. 115–126 (2001)

[12] Sullivan, S., Ponce, J.: Automatic model construction, pose estimation, and object recognition from photographs using triangular splines. IEEE Transactions on Pattern Analysis and Machine Intelligence 20, 1091–1096 (1998)

[13] Ruttle, J., Manzke, M., Dahyot, R.: Smooth kernel density estimate for multiple view reconstruction. In: The 7th European Conference for Visual Media Production (2010)

[14] Exner, D., Bruns, E., Kurz, D., Grundhöfer, A., Bimber, O.: Fast and robust camshift tracking. In: 2010 IEEE Computer Society Conference on Computer Vision and Pattern Recognition Workshops, pp. 9–16 (2010)

[15] Srinivasan, B.V., Hu, Q., Duraiswami, R.: Graphical processors for speeding up kernel machines. In: Workshop on High Performance Analytics - Algorithms, Implementations, and Applications, Siam Conference on Data Mining (2010)

[16] Cootes, T.F., Edwards, G.J., Taylor, C.: Active appearance models. IEEE Transactions on Pattern Analysis and Machine Intelligence, 681–685 (2001)

[17] Lefebvre, S., Hoppe, H.: Parallel controllable texture synthesis. In: ACM SIGGRAPH (2005)

[18] Geusebroek, J.M., Burghouts, G.J., Smeulders, A.W.M.: The amsterdam library of object images. International Journal Computer Vision 61(1), 103–112 (2005)

[19] Kim, D.: 3D Object Reconstruction using Multiple Views. PhD thesis, School of Computer Science and Statistics, Trinity College Dublin, Dublin, Ireland (2011)

[20] Paysan, P., Knothe, R., Amberg, B., Romdhani, S., Vetter, T.: A 3d face model for pose and illumination invariant face recognition. In: The 6th IEEE International Conference on Advanced Video and Signal based Surveillance (AVSS) for Security, Safety and Monitoring in Smart Environments (2009)

Shape Retrieval and Recognition on Mobile Devices

Levente Kovács

Distributed Events Analysis Research Group
Computer and Automation Research Institute, Hungarian Academy of Sciences
Budapest, Hungary
levente.kovacs@sztaki.hu
http://web.eee.sztaki.hu/~kla

Abstract. This paper presents a proof-of-concept shape/contour-based visual recognition and retrieval approach with the main goal of lightweight implementation on mobile devices locally, not relying on network connection or server side processing. In such circumstances the focus needs to be on effectiveness and simplicity, while still preserving high level of functionality (i.e. good recognition). Application areas involve offline object recognition and template matching (e.g. for authorization and blind aid applications), and various object categorizations either in pre-processing or in full local processing.

Keywords: mobile vision, shape recognition, indexing, retrieval.

1 Introduction

As mobile devices proliferate, and include higher processing capabilities, previously workstation-based solutions spread to these devices, either in fully local processing mode, or combined with partial or full server side processing. Targeted areas range from signal processing to high level visual processing tasks. While server side processing can have its benefits - availability of large datapools, of practically unlimited processing power and storage space, etc. -, local processing can also have positive points: low latency, quick responses, non-dependence on network availability or speed, lower costs associated with data transfers and bandwidth. Also, local processing might serve as a pre-processing step of the server-side computations, reducing communication bandwidth and associated latency and cost.

This work presents the first step towards creating a targeted visual recognition solution with local processing on mobile devices, with network and cost non-dependency. The focus is on creating efficient and fast implementations of algorithms, based on the constrained environment (processing power, memory, battery) of mobile devices. The targeted scenario is when we have some a priori information about the possible recognition tasks (e.g. signs, labels, object types) then we can use offline built indexes, upload them to the device, and perform all recognition steps on the device by using these indexes. As a proof-of-concept,

E. Salerno, A.E. Çetin, and O. Salvetti (Eds.): MUSCLE 2011, LNCS 7252, pp. 90–101, 2012.

in this work we present a shape-based visual recognition approach which is run locally on a mobile device, and is able to perform queries against a pre-built compact index containing information about an a-priori shape dataset, with the goal of quickly providing a result for that query, which is the type (class, category, label) of the query object. The presented approach uses a synthetic a-priori shape dataset with versatile content, to provide an example of the possibilities of such a solution.

Traditionally, contours/shape descriptors have been extracted and compared with a series of methods, including Hidden Markov Models [19,2], Scale Invariant Feature points (SIFT) [14], tangent/turning functions [16,13], curvature maps [5], shock graphs [18], Fourier descriptors [4,20], and so on. They all have their benefits and drawbacks, regarding computational complexity, precision capabilities, implementation issues, robustness and scalability. See [15] for one of many comparisons performed between some of these methods.

The works in [19,2] curvature features of contour points are extracted and used to build Hidden Markov Models, and some weighted likelihood discriminator function is used to minimize classification errors between the different models, and good results (64-100% recognition rates) are presented achieved in the case of plane shape classification. In [5] curvature maps are used to compare 3D contours/shapes. In [4,20] Fourier descriptors are used, as probably the most traditional way of representing contour curves, for comparison purposes. In [20] Support Vector Machine based classification and self-organizing maps are both used for contour classification, which results in a robust, yet highly complex and computationally expensive method, resulting in recognition (precision) rates above 60%, and above 80% in most cases. Turning/tangent function based contour description and comparison [16,13] are also used, mostly for comparison purposes, for it being lightweight and fairly easy to implement. These methods work by representing the contours as a function of the local directional angle of the contour points along the whole object, and comparing two such representations.

In this paper we used a modified turning function based contour comparison metric (also used in [12] and presented in detail in [11]). Here, since constraints in mobile computational power, storage space and memory, we use a modified version, presented later.

The novel and interesting elements of the presented approach are the following: providing a proof-of-concept of near-realtime content-based search locally and offline on a mobile device, with local storage, concentrating on speed and effectiveness for low latency, capable of using a number of different a-priori indexes thus providing the possibility of multi-class and multi-feature recognition.

2   Concept, Architecture

The idea of local processing on mobile devices is not new. Yet, until recent years, mobile devices lacked the processing capacity to perform any decent computation offline. Thus most of the higher requirement applications have been built around the Web application and/or cloud based services paradigm. This direction has proved its viability, thousands of applications and services use such architecture to provide server-side processing based consumer interaction. Yet, this paradigm has one flaw: dependence on constant, reliable, cheap (also considering data roaming), high bandwidth network connection. One or two of these parameters might be always reachable, but it is very seldom that all of them are.

Thus, as local computation capabilities of mobile devices rise, offline, or combined offline + online service can be a viable alternative, especially when processing would require a high volume of data propagation. E.g. pre-processing images on the device and not sending it to a server (only processed data) can result in real latency and bandwidth reduction (Fig. 1a).

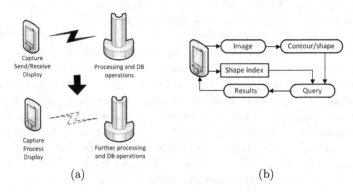

(a)   (b)

Fig. 1. (a) Top: online: capture data on the device, send, wait for results, then receive and display. Bottom: offline: capture, process and display on the device, and only send upstream if necessary. (b) Preferably all processing is performed on the device, without bandwidth, latency and cost issues.

In our case, we are dealing with visual processing, which has always been a computationally intensive and high data volume field. It is for this reason that most approaches and services work with server side processing [7,1], and only a few small apps provide local processing [9]. Recently there has been a new interest in combined local and server side processing apps [17].

The motivation behind this work was to find a solution for the above problems, in the field of visual recognition tasks. In the case of visual recognition, large datasets, a-priori training, high computation complexity are hard to overcome, but there might be situations and applications when local offline processing can be used. One of these areas is on-device shape based recognition of objects, and extraction of shape features as a pre-processing step for server side tasks.

The shape-based processing consists of the following steps (Fig. 1b): loading a prepared shape index file onto the device, capturing or selecting a query shape, performing the query and presenting the result. For the implementation we chose the Android platform[6], for the easy access and availability, and the capability of combining Java and native C++ code in the same application. Development was done using an Android v2.2 emulator, and real tests were performed on Samsung Galaxy 3 (i5800) phone (CPU 667MHz single core, 256MB RAM).

Fig. 2 contains example screenshots of the proof-of-concept application. Query images can be picked from a folder on the device or shot with the camera, then the index file is loaded and the search can be performed. The result (the closest match) is displayed in the textual status line at the top of the window.

2.1   Implementation Issues

During the implementation of the presented algorithmic steps, effectiveness and platform limitations are always an issue to be considered. First, processing capabilities of mobile devices are rising, but there are still quite a number of lower capacity models. Second, on most platforms there are limitations to be considered, e.g. maximum heap/stack sizes, memory capacity, storage space limitations, etc. Some of the computational issues (e.g. Java applications' memory limitations) can be somewhat alleviated in the case of Android, since there is a native part of the SDK available, thus Java code can be combined with C++ in the same application through Java/JNI interfaces. In the case of the proof-of-concept test application, we used the same approach (Fig. 3): we put as much of the algorithms into native code as possible, with the goal of higher speed and effectiveness.

Also, for memory and storage considerations, the dataset used for the building of the index is not transferred to the mobile device, only the index is uploaded, which only contains the shape information of the nodes and an identifier to be able to provide answers to queries.

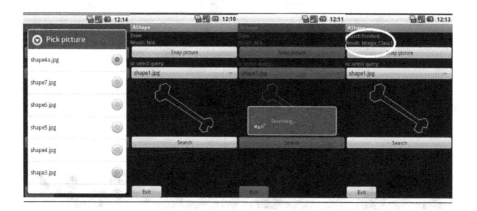

Fig. 2. Screencaps of the proof-of-concept app from the Android device emulator. From left to right: picking a stored image for query; main interface (showing picked or captured query); performing the search; displaying the result.

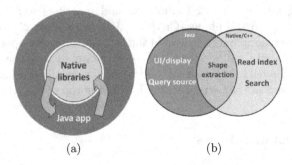

(a)   (b)

Fig. 3. (a) For speed considerations and memory requirements, most of the algorithms run transparently in native C++ on the Android device, used as external libraries from the Java application. (b) High level grouping of algorithmic elements based on their implementation.

3   Indexing, Dataset

The basis of the test dataset used is the one used in [2], extended with other synthetic shape classes[19]. The original number of 673 shapes in 30 classes was reduced to 534 shapes after filtering out very similar shapes during the index building process. Fig. 4 shows some examples of shapes from each of the classes. In-class variations include different distortions of the same shape (Fig. 5). Table 1 shows the number of shapes in each class after the indexing is performed.

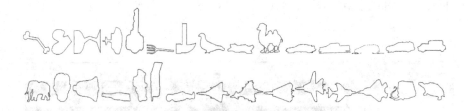

Fig. 4. One example from each of the used synthetic shape classes. Each class contains various versions of each shape.

Fig. 5. Example for in-class shape variations

Table 1. Number of elements in each of the used shape classes

class number	1	2	3	4	5	6	7	8	9	10	11	12	13	14	15
nr. of elements	20	20	20	40	20	20	20	12	11	11	30	30	30	30	12

class number	16	17	18	19	20	21	22	23	24	25	26	27	28	29	30
nr. of elements	12	8	20	20	20	20	20	17	2	22	7	3	20	7	10

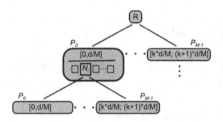

Fig. 6. Structure of an index tree

To facilitate effective searching in the retrieval/recognition phase, an a-priori index structure needs to be built which can be loaded on the device, to be available when the application starts. The index should be able to handle shape features. Obviously, a lot of different tree structures are available for representing relations between all kinds of objects, including classic ones like M, B, R, TV, etc. and complex ones for multidimensional data like BVH trees, KD-trees, and so on. In choosing an index structure, especially in the case of high performance in a low computation capacity environments, easy handling, low memory footprint and fast searching are the main desired parameters, while time to build the index are only marginally important. Our previous works where we needed realtime search capabilities[12] led us to choose a modified BK-tree[3] solution. One of the main benefits of the BK-tree-based approach is that after a proper index building, during the search large parts of the index tree can be dropped even from the first level, and there is no need to revisit other branches on the same level again. This makes searching through the tree a fast process.

The index structure is a variation of BK-trees introduced earlier[12,11] (Fig. 6). Essentially the index trees are representations of point distributions in discrete metric spaces. For classical string matching purposes, the tree is built so as to have each subtree contain sets of strings that are at the same distance from the subtree's root, i.e. for all e leaves below sub-root r the $d(e, r) = \varepsilon$ is constant. In our case, the modifications include a). using a tree structure where the nodes contain not only an identifier, but also the descriptor data associated to the node, which makes searching and result display a fast process, and b). using a structure that contains nodes that can have an arbitrary number of children (N), where the leaves below each child contain elements for which the distance d falls in a difference interval: $d(e, r) \in [\varepsilon_i; \varepsilon_{i+1})$ (where $i \in [0, N] \cap \mathbb{N}$). The distance intervals in the child nodes (denoted by $\varepsilon_i, \varepsilon_{i+1}$ above) depend on the maximum error E_{max} that the distance metric can have, more specifically, $\|\varepsilon_{i+1} - \varepsilon_i\| = E_{max}/N$, thus the intervals are linearly divided buckets.

The result of the indexing process is a binary file that contains a compact representation of the built index tree structure. When searching the index, this binary file is loaded, from which the tree structure can be replicated.

4   Retrieval, Recognition

For the dataset contour point data is already available, thus they are used (after a noise filtering step to produce smoother contours) during the index building process.

To localize objects in the query image, we use a contour point detection method [10] based on the Harris corner detection [8] as a starting point. The traditional approach only emphasizes corners in the image, but for contour point detection we need both corner and edge points to be extracted, therefore we need to modify the original characteristic function.

The eigenvalues of the so called Harris matrix[8] M (denoted by λ_1 and λ_2) will be proportional to the principal curvature of the local autocorrelation function and separate three kinds of regions: both of them are large in corner regions, only one of them is large in edge regions and both of them are small in homogeneous (flat) regions. In the modified detector points are then detected as the local maxima of $L = \max(\lambda_1, \lambda_2)$ around pixel (x_i, y_i):

$$(x_i^\star, y_i^\star) = \underset{(x_i, y_i) \in b}{\operatorname{argmax}} \{L(x_i, y_i)\} \, , \tag{1}$$

and the set of detected $(x_i^\star, y_i^\star)$ points will be used as corner/contour points. Fig. 7 shows examples to compare the Harris detector with the modified detector, while Fig. 8 shows an example contour detection.

As a metric for comparing different extracted contours/shapes, we used a turning function approach[12,11]. The reason for choosing such an approach is the lightweight nature of the method, and its tolerance against rotation, scale and noise/changes in the shapes. Fig. 9 shows an example for internal representation of a shape.

In recognizing shapes, the result of a query is the best match. Yet, to evaluate the metric and the comparison capabilities of the approach, we included the results of retrieval tests, and the first 5 results for 7 different queries are displayed as examples in Fig. 12. The results show the viability of the approach.

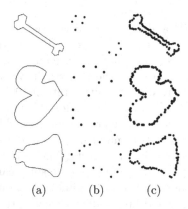

Fig. 7. Contour point detection. (a): samples; (b): original Harris corner detector; (c): used detector[10].

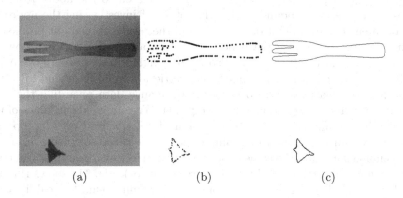

Fig. 8. Sample input: (a) image; (b) detected points; (c) produced contour.

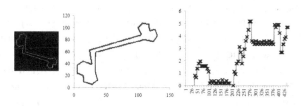

Fig. 9. Sample for input image (left), extracted contour (middle), internal turning representation (right)

4.1   Evaluation

We investigated the visual and numerical performance of the above presented approach on the outlines dataset. First, recognition/retrieval capabilities were investigated for numerical results, which are presented in Fig. 10.

For all queries and retrieval tests, the first result was always inline with the query, which means we received a recognition rate of 100% (of course in the case when queries were belonging to one of the indexed classes).

We investigated the relation between retrieval threshold bounds (affecting the number of returned results) and the number (percentage) of visited nodes in the index tree (Fig. 10a), which directly affects time performance. The expected result was that allowing a higher number of results causes large increase in the number of visited nodes (and required time). We also investigated how the change in the number of visited nodes affects recall rates (Fig. 10b). Here, the expected outcome - which was also the produced result - was that an increase in visited nodes results in an increase of recall values (note: recall is the number of relevant results with relation to the total number of possible relevant dataset elements).

Fig. 10c shows precision-recall (P/R) curves related to the above tests (20 queries with 7 retrieval bounds) with the intent of investigating the approach from a content retrieval point of view. Overall the results turned out promising: the first returns (at low retrieval bounds) which are used for recognition are always $P = 1$, and recall rates are also good, ranging from 40 to 100%. Additionally, we included in Fig. 12 some visual retrieval results to extend Fig. 10c.

We also ran measurements to gather performance data regarding the time required to produce recognition results (Fig. 11). We compared run times of the same algorithm running natively (C++) on an Intel 1.3GHz Core2 laptop processor, in an Android emulator running on the same processor (using Java and C++ combination), and the same code running natively on a Samsung Galaxy 3 Android phone (667MHz ARM architecture processor). Fig. 11a shows running times for all the above queries on a laptop. The default bound for real life tests is bound nr. 2, which means that generally results are provided below 500ms. Fig. 11b shows average running times for all the queries on all 3 environments. As it is shown, there is an order of magnitude difference between the PC and the 667MHz mobile device results. What this means in practice is, that for a lower number of objects (below 500), and a lower number of categories (practically below 20) close-to-realtime results can be achieved. Application areas in such scenarios include face recognition, iris recognition, fingerprint recognition (for identification and authorization purposes), or blind aid applications (currency/coin recognition, street sign recognition, etc.), all working reliably without constant network connection. Also, today we already have mobile devices (phones and tablets) with processors up to 1GHz in clock speed (some with multiple cores), which can cause significant increase in processing speed. Finally, we can conclude that local offline processing on mobile devices is a practically viable approach and good results can be expected.

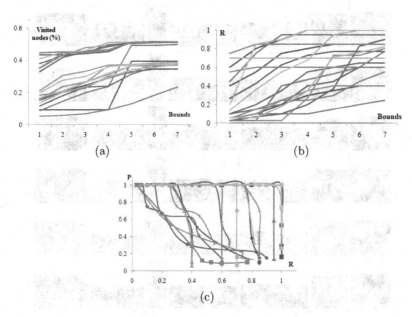

Fig. 10. Evaluation of retrieval (performance). (a): Visited nodes of the index tree (in %) for 20 different query curves based on 7 decreasing retrieval bounds (horizontal axis). (b): Recall rates associated to the same queries and retrieval bounds as in (a). Meaning: as more nodes are visited in the tree, Recall rates increase accordingly. (c): Precision/Recall curves for 20 different queries, each for 7 different retrieval bounds.

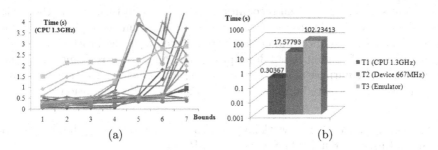

Fig. 11. Evaluation of retrieval (time). (a) Retrieval time curves for the same 20 queries and 7 increasing retrieval bounds (default is bound nr. 2) related to Fig. 10 - when run on a notebook Core2 CPU 1.3GHz. (b) Average time comparisons for the same set of runs on a 1.3GHz Core2 CPU, a Samsung Galaxy 3 (667MHz) and an Android Emulator (running on the same 1.3GHz CPU).

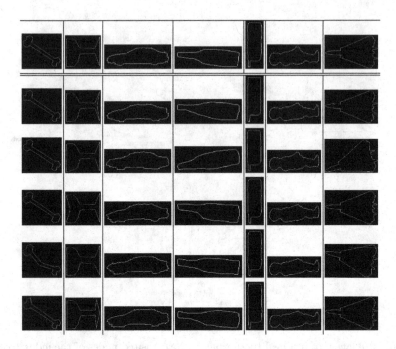

Fig. 12. Samples for retrievals for different query images (top row) and the first 5 results (in columns)

5   Conclusions

In this paper we presented an approach for a simple and fast shape-based retrieval and recognition solution for local offline processing on mobile devices. The purpose of such methods is to provide a basis for reliable, non-network-dependent solutions for visual recognition tasks. These in turn can be the basis of local authorization/login/authentication solutions, blind aid applications (sign recognition, cash classification, etc.). Currently we are in the process of providing a larger set of content based descriptors that can be used realtime on mobile devices, with the purpose of local feature extraction and categorization of a small set of objects for targeted applications (product and sign recognition).

Acknowledgements. This work has been partially supported by the Hungarian Scientific Research Fund (OTKA) grant nr. PD83438.

References

1. Aurasma: Aurasma Augmented Reality Platform, http://www.aurasma.com
2. Bicego, M., Murino, V.: Investigating Hidden Markov Models' capabilities in 2D shape classification. IEEE Tr. on Pattern Recognition and Machine Intelligence 26(2), 281–286 (2004)

3. Burkhard, W., Keller, R.: Some approaches to best-match file searching. Communications of the ACM 16, 230–236 (1973)
4. Frejlichowski, D.: An Algorithm for Binary Contour Objects Representation and Recognition. In: Campilho, A., Kamel, M.S. (eds.) ICIAR 2008. LNCS, vol. 5112, pp. 537–546. Springer, Heidelberg (2008)
5. Gatzke, T., Garland, M.: Curvature maps for local shape comparison. In: Proc. of Shape Modeling and Applications, pp. 244–253 (2005)
6. Google: Google Android SDK, http://developer.android.com
7. Google: Google Goggles, http://www.google.com/mobile/goggles
8. Harris, C., Stephens, M.: A combined corner and edge detector. In: Proc. of the 4th Alvey Vision Conference, pp. 147–151 (1988)
9. Ipplex: LookTel Money Reader, http://www.looktel.com/products
10. Kovacs, A., Sziranyi, T.: High Definition Feature Map for GVF Snake by Using Harris Function. In: Blanc-Talon, J., Bone, D., Philips, W., Popescu, D., Scheunders, P. (eds.) ACIVS 2010, Part I. LNCS, vol. 6474, pp. 163–172. Springer, Heidelberg (2010)
11. Kovács, L.: Contour Based Shape Retrieval. In: Bebis, G., Boyle, R., Parvin, B., Koracin, D., Chung, R., Hammound, R., Hussain, M., Kar-Han, T., Crawfis, R., Thalmann, D., Kao, D., Avila, L. (eds.) ISVC 2010, Part III. LNCS, vol. 6455, pp. 59–68. Springer, Heidelberg (2010)
12. Kovács, L., Utasi, A.: Shape and motion fused multiple flying target recognition and tracking. In: Proc. of Automatic Target Recognition XX, SPIE Defense, Security and Sensing, vol. 7696, pp. 769605-1–12 (2010)
13. Latecki, L.J., Lakamper, R.: Application of planar shape comparison to object retrieval in image databases. Pattern Recognition 35(1), 15–29 (2002)
14. Lowe, D.G.: Object recognition from local scale-invariant features. In: ICCV, pp. 1150–1157 (1999)
15. Rosenhahn, B., Brox, T., Cremers, D., Seidel, H.-P.: A Comparison of Shape Matching Methods for Contour Based Pose Estimation. In: Reulke, R., Eckardt, U., Flach, B., Knauer, U., Polthier, K. (eds.) IWCIA 2006. LNCS, vol. 4040, pp. 263–276. Springer, Heidelberg (2006)
16. Scassellati, B., Alexopoulos, S., Flickner, M.: Retrieving images by 2D shape: a comparison of computation methods with perceptual judgements. In: Proc. of SPIE Storage and Retrieval for Image and Video Databases II, vol. 2185, pp. 2–14 (1994)
17. Schroth, G., Huitl, R., Chen, D., Abu-Alqumsan, M., Al-Nuaimi, A., Steinbach, E.: Mobile visual location recognition. IEEE Signal Processing Magazine 28(4), 77–89 (2011)
18. Sebastian, T., Klein, P.N., Kimia, B.B.: Recognition of shapes by editing their shock graphs 26(5), 550–571 (2004)
19. Thakoor, N., Gao, J., Jung, S.: Hidden Markov Model-based weighted likelihood discriminant for 2D shape classification. IEEE Tr. on Image Processing 16(11), 2707–2719 (2007)
20. Wong, W.T., Shih, F.Y., Liu, J.: Shape-based image retrieval using support vector machines, fourier descriptors and self-organizing maps. Intl. Journal of Information Sciences 177(8), 1878–1891 (2007)

Directionally Selective Fractional Wavelet Transform Using a 2-D Non-separable Unbalanced Lifting Structure

Furkan Keskin and A. Enis Çetin

Department of Electrical and Electronics Engineering,
Bilkent University, Bilkent, 06800, Ankara, Turkey
keskin@ee.bilkent.edu.tr
cetin@bilkent.edu.tr

Abstract. In this paper, we extend the recently introduced concept of fractional wavelet transform to obtain directional subbands of an image. Fractional wavelet decomposition is based on two-channel unbalanced lifting structures whereby it is possible to decompose a given discrete-time signal $x[n]$ sampled with period T into two sub-signals $x_1[n]$ and $x_2[n]$ whose average sampling periods are pT and qT, respectively. Fractions p and q are rational numbers satisfying the condition: $1/p + 1/q = 1$. Filters used in the lifting structure are designed using the Lagrange interpolation formula. 2-d separable and non-separable extensions of the proposed fractional wavelet transform are developed. Using a non-separable unbalanced lifting structure, directional subimages for five different directions are obtained.

Keywords: Lifting, wavelet transform, multirate signal processing.

1 Introduction

Lifting structures provide computationally efficient implementation of the wavelet transform [1,2,3,4,5,6,7] and they found applications in image and video coding and signal and image analysis applications [8,9].

New unbalanced wavelet lifting structures producing directional decomposition of the input image are introduced in this article. In standard lifting structures the input signal is first decomposed into even and odd indexed samples using the lazy wavelet transform. In the recently proposed unbalanced lifting structure, a structure similar to the lazy filterbank forwards every p^{th} sample of the original signal to the upper-branch and remaining $p - 1$ samples out of p samples go to the lower branch [10]. Discrete-time update and prediction filters interconnect the upper and lower branches, respectively. Discrete-time filters operating in the lower sampling rates are designed using the Lagrange interpolation formula which is also used in many filterbank designs including our filterbanks, Smith-Barnwell and Daubechies filterbanks. In this way, two sub-signals with different sampling rates are obtained from the original signal. When the input

E. Salerno, A.E. Çetin, and O. Salvetti (Eds.): MUSCLE 2011, LNCS 7252, pp. 102–113, 2012.
© Springer-Verlag Berlin Heidelberg 2012

signal has a sampling rate of T the upper- and lower- branches of the unbalanced lifting filterbank have sampling rates of pT and $Tp/(p-1)$, respectively. In standard balanced lifting the sampling periods of upper and lower branches are the same: $2T$.

The unbalanced lifting decomposition can be easily generalized to other sampling strategies in which the upper-branch has a sampling rate of pT and the lower-branch has a sampling rate of qT with the property that

$$\frac{1}{p} + \frac{1}{q} = 1 \tag{1}$$

Perfect reconstruction can be easily achieved by changing the signs of the filters in the reconstruction part of the ordinary balanced lifting structures.

In Section 2, an example filterbank design with $p = 3 : 1$ and $q = 3 : 2$ is presented. In Section 3, 2-D separable filterbank design examples are presented. Non-separable 2-D extension of the unbalanced lifting structure resulting in directional subbands is developed in Section 4.

2 Unbalanced Lazy Filterbank and Lifting Structures

An unbalanced lazy filterbank for $p = 3 : 1$ and $q = 3 : 2$ is shown in Figure 2. In the upper-branch a regular downsampling block by a factor of three is used. In the lower-branch the downsampling block for $q = 3 : 2$ is used. In Figure 2, the signal $x[n]$ is fed to the lazy filterbank and outputs of downsampling blocks are shown. Every 3rd sample of the original signal goes to the upper-branch and remaining samples appear in the lower branch.

We describe the update and the prediction filter design for the unbalanced lifting structure in the following subsection.

2.1 Update Filter

In Figure 1 an unbalanced lifting structure is shown with downsampling ratios $3 : 1$ and $3 : 2$ in the upper and lower branches, respectively. Similar to the regular balanced lifting filterbank case, the upper branch sample d can be estimated using the neighboring lower branch samples $b, c, e,$ and f and an estimate of the sample d is given as follows:

$$d' = \frac{b + 2c + 2e + f}{6}, \tag{2}$$

which is the output of the update filter linking the lower branch to the upper branch. Since samples c and e are closer to the sample d compared to f and b more weight is given to the samples c and e. The sample d and the output of the filter is linearly combined and the updated sample is obtained as follows

$$\hat{d} = \frac{2d'}{3} + \frac{d}{3} \tag{3}$$

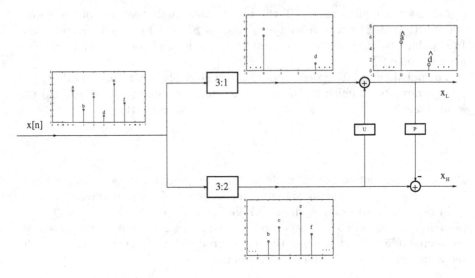

Fig. 1. Unbalanced lifting scheme with $p = 3 : 1$ and $q = 3 : 2$ downsampling ratios

Since the downsampling is by a factor of three in the upper-branch original signal must be filtered with a low-pass filter with a cut-off frequency of $\pi/3$. We could have used $\hat{d} = \frac{d'}{2} + \frac{d}{2}$ but this does not produce as good frequency response as (3). Therefore, the upper branch output sample is given by

$$\hat{d} = \frac{2b}{18} + \frac{4c}{18} + \frac{6d}{18} + \frac{4e}{18} + \frac{2f}{18} \tag{4}$$

The corresponding filter coefficients are given as follows

$$h_l[n] = \{\frac{2}{18}, \frac{4}{18}, \frac{6}{18}, \frac{4}{18}, \frac{2}{18}\} \tag{5}$$

The frequency response of this filter is plotted in [10]. The cut-off frequency of this filter is $\pi/3$.

Other samples of the upper branch of the filterbank are smoothed in a similar manner. In Equation (5), each sample of the upper branch sub-signal is updated using four neighboring samples of the lower branch. It is possible to smooth the samples of the upper branch further by using more samples from the lower branch without effecting the perfect reconstruction capability of the filterbank. In this case, Lagrange interpolation formula or other interpolation methods can be used to determine the update filter coefficients. In general,

$$x_L[n] = \frac{1}{9}x[3n-2] + \frac{2}{9}x[3n-1] + \frac{3}{9}x[3n] + \frac{2}{9}x[3n+1] + \frac{1}{9}x[3n+2] \tag{6}$$

where $x[n]$ is the original input signal to the filterbank.

2.2   Prediction Filter

Samples of the lower branch are estimated from the upper branch and difference is transmitted to the receiver. Lower branch samples can be predicted from the upper branch using the updated samples $\hat{a}, \hat{d}, \hat{g}, \ldots$ (see Figure 1). The prediction filter can be as simple as the identity operator selecting the nearest upper branch sample as an estimate of the lower branch sample:

$$x_H[0] = b - \hat{a}, x_H[1] = c - \hat{d}, \ldots \tag{7}$$

where the subscript H indicates that $x_H[n]$ is a high-band sub-signal. Other samples of $x_H[n]$ can be determined in a similar manner. Although the above prediction strategy is very simple and computationally efficient, the above predictor is not a good estimator. We can use Lagrange interpolation and obtain:

$$x_H[0] = b - (2\hat{a} + \hat{d})/3, x_H[1] = c - (2\hat{d} + \hat{a})/3, \ldots \tag{8}$$

In general,

$$x_H[n] = \begin{cases} x\left[\frac{3n+2}{2}\right] - \frac{2x_U\left[\frac{n}{2}\right] + x_U\left[\frac{n}{2}+1\right]}{3}, & \text{n is even} \\ x\left[\frac{3n+1}{2}\right] - \frac{2x_U\left[\frac{n+1}{2}\right] + x_U\left[\frac{n-1}{2}\right]}{3}, & \text{n is odd} \end{cases} \tag{9}$$

where $x[n]$ is the original input signal to the filterbank.

To determine $x_H[0]$ the sample b is estimated using the two nearest upper branch samples $\hat{a}$ and $\hat{d}$. Let $P(t)$ be the Lagrange interpolator based on the samples $\hat{a}$ and $\hat{d}$:

$$P(t) = \hat{a}\,\ell_o(t) + \hat{d}\,\ell_1(t) \tag{10}$$

where $\ell_i(t)$ are the Lagrange basis polynomials. The function $P(t)$ is constructed using the input data pairs $(\hat{a}, 0)$ and $(\hat{d}, 3T)$. Since the sample b occurs at $t = T$, $\ell_o(T) = 2/3$ and $\ell_1(T) = 1/3$. Therefore the Lagrange interpolation gives more weight to the sample $\hat{a}$ because it is nearer to the sample b compared to $\hat{d}$ to determine $x_H[0]$ in (8). Similarly, more weight is given to the sample $\hat{d}$ because it is nearer to the sample c compared to $\hat{a}$ to determine $x_H[1]$ etc. As it can be seen from the above equations the predictor is a time varying filter. The predictor can even be an adaptive LMS-type filter as described in references [4,11,12]. The adaptive prediction tries to remove as much information as possible from the lower-branch using $x_H[n]$ samples. In this case the computational complexity is higher than the predictors in (7) and (8). Other samples of $x_L[n]$ can be also used by the predictor filter. In this case higher order Lagrange interpolation formula needs to be used [4,13].

It is trivial to design the reconstruction filterbank as in regular lifting structures. At the reconstruction stage signs of filters U and P are changed and sub-signal samples are realigned to obtain the original signal $x[n]$.

Designing unbalanced lifting structures with different p and q values is also possible by following the abovementioned design strategy. For instance, an unbalanced lifting scheme with $p = 3 : 2$ and $q = 3 : 1$ can be constructed by

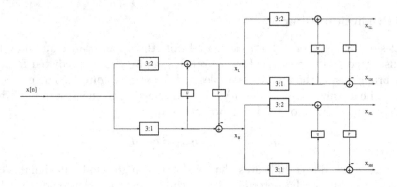

Fig. 2. Extension of the unbalanced lifting structure with p=3:2 and q=3:1 to 2-D

determining the update and prediction filter coefficients according to distances between samples. It should be noted that low-pass filtered upper-branch signal is not uniformly sampled in this case. Hence, update and prediction filters are designed after transforming the upper branch samples to a uniform grid. Then, update and prediction operations can be carried out in a similar manner.

3   Extension for Two-Dimensional Signals

In Figure 2, the separable filterbank structure for the unbalanced lifting wavelet transform with $p = 3 : 2$ and $q = 3 : 1$ is given. The input image is first downsampled in the horizontal direction. Then the update and prediction filters are applied to the downsampled images. The intermediate output signals x_L and x_H are then downsampled in the vertical direction and the update and prediction filters are applied again.

As an example, the image given at Figure 3 (left) is fractional wavelet transformed using the scheme presented in Figure 2. The transformed image is given at Figure 3 (right). As another example, the image given at Figure 4 is wavelet transformed in Figure 1 using the fractional wavelet transform with $p = 3 : 1$ and $q = 3 : 2$ which is a 2-D extended version of the filterbank described in Section 2.

4   2-D Directionally Selective Non-separable Unbalanced Filterbank Structure

The 2-D non-separable extension of the unbalanced lifting wavelet transform is also possible. In Section 3, separable filterbank structure with $p = 3 : 2$ and $q = 3 : 1$ is proposed for two-dimensional signals. In this section, we design a non-separable lifting structure for 2-D signals with $p = 9 : 1$ and $q = 9 : 8$, where p and q denote spatial sampling rates. Figure 5 shows the non-separable structure having one upper branch and five lower branches, each corresponding to a prediction filter designed to reveal image edges in a specific direction.

Fig. 3. Example image (left) and 2-D unbalanced lifting wavelet transformed image with p=3:2 and q=3:1 (right). Highband subimages are amplified by a factor of five.

Fig. 4. 2-D unbalanced lifting wavelet transformed image with p=3:1 and q=3:2. High-band subimages are amplified by a factor of five.

In Figure 5, the upper branch has a downsampling ratio of 9 : 1 and the lower branch has a downsampling ratio of 9 : 8. First, a 2-D update filter is used to smooth the upper branch samples so that a low-resolution image $x_L[n_1, n_2]$ is obtained. Afterwards, 2-D prediction filters are used to obtain the directional subbands of the input image in five different directions, which are 0, 26.5, 45, 63.4 and 90 degrees with respect to the horizontal axis.

In Figure 5, pixels of an image are shown. The pixel marked e can be estimated from the neighboring lower branch samples $a, b, c, d, f, g, h,$ and i as follows:

$$e' = \frac{a + c + g + i + \sqrt{2}b + \sqrt{2}f + \sqrt{2}h + \sqrt{2}d}{4\sqrt{2} + 4} \tag{11}$$

The weights of the neighboring samples in estimating the sample e are determined based on their geometric distance to the sample e. The estimate obtained in Equation (11) represents the output of the update filter and is linearly combined with the upper branch sample e to obtain the updated sample corresponding to e:

$$\hat{e} = w_1 e' + w_2 e \tag{12}$$

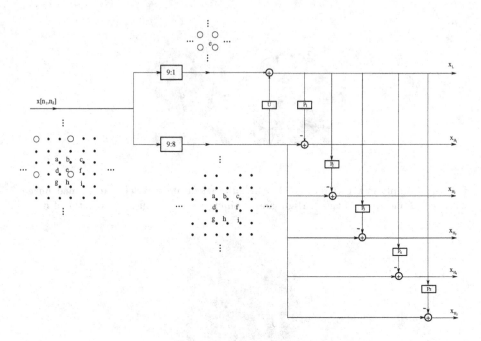

Fig. 5. 2-D non-separable extension of the unbalanced lifting structure with $p = 9 : 1$ and $q = 9 : 8$

where w_1 and w_2 are weights which can be determined as in Section 4 such that the effective filter is a low-pass filter with cut-off frequency $w = \pi/3$ in both horizontal and vertical directions.

Similar to the unbalanced lifting structure for 1-D signals, the original 2-D signal must be filtered using a low-pass filter to avoid aliasing. The spatial downsampling ratio is $9 : 1$ in the upper branch, which corresponds to a downsampling ratio of $3 : 1$ in both horizontal and vertical directions. When $w_1 = 8/9$ and $w_2 = 1/9$, the upper branch output sample is computed as

$$\hat{e} = \frac{2}{9(\sqrt{2}+1)}(a+c+g+i) + \frac{2\sqrt{2}}{9(\sqrt{2}+1)}(b+f+h+d) + \frac{1}{9}e \quad (13)$$

The coefficients of the filter that yields the upper branch samples from the original signal are given by

$$h_l[n_1, n_2] = \begin{bmatrix} \alpha & \beta & \alpha \\ \beta & \gamma & \beta \\ \alpha & \beta & \alpha \end{bmatrix} \quad (14)$$

where $\alpha = \frac{2}{9(\sqrt{2}+1)}$, $\beta = \frac{2\sqrt{2}}{9(\sqrt{2}+1)}$, and $\gamma = 1/9$.

Frequency response of this symmetric filter is shown in Figure 6.

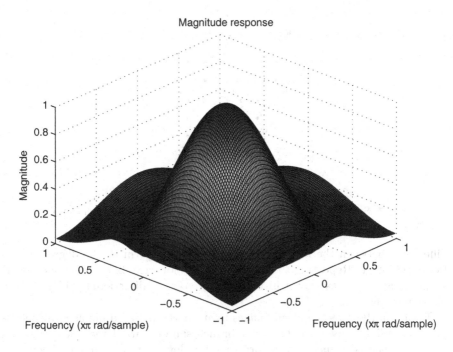

Magnitude response

Fig. 6. Frequency response of the filter in Equation (14)

In general, the upper branch samples are given by

$$x_L[n_1, n_2] = \gamma x[3n_1 - 2, 3n_2 - 2] + \alpha(x[3n_1 - 3, 3n_2 - 3] +$$
$$x[3n_1 - 3, 3n_2 - 1] + x[3n_1 - 1, 3n_2 - 3] + x[3n_1 - 1, 3n_2 - 1]) +$$
$$\beta(x[3n_1 - 3, 3n_2 - 2] + x[3n_1 - 1, 3n_2 - 2] + \qquad (15)$$
$$x[3n_1 - 2, 3n_2 - 3] + x[3n_1 - 2, 3n_2 - 1])$$

where $x[n_1, n_2]$ is the original input image to the filterbank.

Prediction filters P_1, P_2, P_3, P_4, and P_5 are designed to find estimates of the lower branch samples using the closest upper branch samples in five different directions, which are 0, 26.5, 45, 63.4 and 90 degrees, respectively, with respect to the horizontal axis. Figure 7 shows the upper branch samples $a_{i,j}$ and lower branch samples. The samples y_1 and y_2 are estimated by P_1 using the updated upper branch samples $a_{3,1}$ and $a_{3,2}$, and 0 degree high subband of the input image is constructed. The sample v is estimated by P_2 using the upper branch samples $a_{3,1}$ and $a_{2,3}$, and the differences going into the second lower branch output channel make up the 26.5 degrees directional subband image. For the 45 degree diagonal subband, the samples w_1 and w_2 are predicted by using the upper branch samples $a_{3,1}$ and $a_{2,2}$. P_4 uses $a_{3,1}$ and $a_{1,2}$ samples to obtain the predictor for the sample u, resulting in 63.4 degrees directional subband image. Finally, the prediction filter P_5 produces estimates of the lower branch samples z_1 and z_2 using the upper branch samples $a_{3,1}$ and $a_{2,1}$, and the

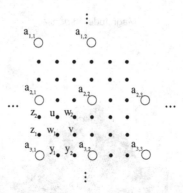

Fig. 7. Spatial organization of upper branch and lower branch samples

differences between the actual values and estimates form the 90 degrees directional subimage. Repeating these computations for each 3x3 subregion of a 2-D signal or an image as shown in Figure 7, we obtain directional subbands in a non-separable manner.

To obtain the directional subimages by using the updated upper branch samples, the weights given to the upper branch samples are determined based on their geometric distance to the corresponding lower branch sample. Therefore, the closest neighboring upper branch sample has the weight 2/3 and the other sample gets the weight 1/3. For instance, for prediction of the sample v, $a_{3,1}$ is assigned 2/3 weight and $a_{2,3}$ is assigned 1/3 weight. Other lower branch samples can be predicted in a similar manner.

In general, directional subimages are given by the following equations:

$$x_{H_1}[n_1,n_2] = \begin{cases} x[3n_1 - 2, \frac{3n_2}{2}] - \frac{2x_L[n_1,\frac{n_2}{2}+1]+x_L[n_1,\frac{n_2}{2}]}{3}, & n_2 \text{ is even} \\ x[3n_1 - 2, \frac{3n_2+1}{2}] - \frac{2x_L[n_1,\frac{n_2+1}{2}]+x_L[n_1,\frac{n_2+3}{2}]}{3}, & n_2 \text{ is odd} \end{cases} \quad (16)$$

$$x_{H_2}[n_1,n_2] = x[3n_1, 3n_2] - \frac{2x_L[n_1 + 1, n_2] + x_L[n_1, n_2 + 2]}{3} \quad (17)$$

$$x_{H_3}[n_1,n_2] = \begin{cases} x[3n_1, \frac{3n_2+1}{2}] - \frac{2x_L[n_1+1,\frac{n_2+1}{2}]+x_L[n_1,\frac{n_2+3}{2}]}{3}, & n_2 \text{ is odd} \\ x[3n_1 - 1, \frac{3n_2}{2}] - \frac{2x_L[n_1,\frac{n_2}{2}+1]+x_L[n_1+1,\frac{n_2}{2}]}{3}, & n_2 \text{ is even} \end{cases} \quad (18)$$

$$x_{H_4}[n_1,n_2] = x[3n_1 - 1, 3n_2 - 1] - \frac{2x_L[n_1 + 1, n_2] + x_L[n_1 - 1, n_2 + 1]}{3} \quad (19)$$

$$x_{H_5}[n_1,n_2] = \begin{cases} x[\frac{3n_1}{2}, 3n_2 - 2] - \frac{2x_L[\frac{n_1}{2}+1,n_2]+x_L[\frac{n_1}{2},n_2]}{3}, & n_1 \text{ is even} \\ x[\frac{3n_1+1}{2}, 3n_2 - 2] - \frac{2x_L[\frac{n_1+1}{2},n_2]+x_L[\frac{n_1+3}{2},n_2]}{3}, & n_1 \text{ is odd} \end{cases} \quad (20)$$

It is also possible to use the four closest neighboring upper branch samples in the desired direction instead of the two closest samples to estimate the lower branch sample. Then, the weights of the updated upper branch samples are assigned accordingly. More robust estimates of the lower branch samples can be achieved by adopting this prediction strategy. Other possibility is to use the adaptive filters in [6] and [7] for predicting the lower branch samples.

Figure 8 shows the directionally transformed image using the lifting scheme presented in Figure 5. Directional subband images reveal directional edges present in the input image. Non-separable unbalanced lifting structure provides directional edge information that can not be obtained by employing the separable filterbank structure. Figure 9 (left) shows another example image and Figure

Fig. 8. 2-D unbalanced directional lifting wavelet transformed image with p = 9:1 and q = 9:8. Highband subimages are amplified by a factor of five

9 (right) shows the directional wavelet transformed subimages using the non-separable unbalanced lifting scheme depicted in Figure 5. As observed from Figure 9, only the lines orthogonal to the direction in which the desired egdes are found are not highlighted in directional subband images. For instance, in 63.4 degree directional subimage, the line that is in approximately 63.4 degree orientation with respect to the horizontal axis has the same gray value as the background, which indicates that there is almost no edge on this line in 63.4 degree direction with respect to the horizontal axis. Similarly, it is evident in 45 degree diagonal subband image that almost no directional edges are found on the line that makes an angle of 45 degree with the horizontal axis.

There are other possible lifting schemes than the one proposed in this paper for directionally selective fractional wavelet decomposition of images in a non-separable manner. An alternative scheme is a 2-D non-separable unbalanced lifting structure where the spatial downsampling ratios are taken to be $p = 16 : 1$ and $q = 16 : 15$ for upper and lower branches, respectively. In this scheme directional subband images can be obtained for more than five directions by adding to the lifting structure several prediction filters tailored to desired directions. In this way, we have a more sophisticated directional wavelet decomposition with more detail about different angles while obtaining a lower resolution low-pass filtered subimage $x_L[n_1, n_2]$. Another possibility for an unbalanced lifting scheme is to employ a downsampling and prediction strategy such that directions ranging from 90 degree to 180 degree are also included in the directional decomposition of the input image.

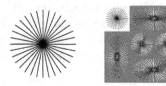

Fig. 9. Example image 2 (left) and 2-D unbalanced directional lifting wavelet transformed image with p = 9:1 and q = 9:8 (right). Highband subimages are amplified by a factor of five.

5   Conclusion

In this article, directionally selective fractional wavelet transform (FWT) methods are developed using unbalanced lifting structures. Sub-images have different average sampling rates.

2-D separable extensions carry out horizontal and vertical processing in different steps of the lifting transform. 2-D non-separable lifting structures provide unbalanced directional decomposition of the input image, leading to many different directional subbands. Image transformation examples are presented for both separable and non-separable cases. The FWT method can be easily extended to higher dimensions.

References

1. Mallat, S.: A Wavelet Tour of Signal Processing, 2nd edn. (Wavelet Analysis & Its Applications). Academic Press (1999)
2. Sweldens, W.: The lifting scheme: A custom-design construction of biorthogonal wavelets. Applied and Computational Harmonic Analysis 3, 186–200 (1996)
3. Daubechies, I., Sweldens, W.: Factoring wavelet transforms into lifting steps. Journal of Fourier Analysis and Applications 4, 247–269 (1998)
4. Kim, C.W., Ansari, R., Cetin, A.E.: A class of linear-phase regular biorthogonal wavelets. In: IEEE International Conference on Acoustics, Speech, and Signal Processing, vol. 4, pp. 673–676 (1992)
5. Hampson, F.J., Pesquet, J.C.: A nonlinear subband decomposition with perfect reconstruction. In: IEEE International Conference on Acoustics, Speech, and Signal Processing, vol. 3, pp. 1523–1526 (1996)

6. Gerek, O.N., Cetin, A.E.: Adaptive polyphase subband decomposition structures for image compression. IEEE Transactions on Image Processing 9, 1649–1660 (2000)
7. Gerek, O.N., Cetin, A.E.: A 2-d orientation-adaptive prediction filter in lifting structures for image coding. IEEE Transactions on Image Processing 15, 106–111 (2006)
8. Le Pennec, E., Mallat, S.: Image compression with geometrical wavelets. In: International Conference on Image Processing, vol. 1, pp. 661–664 (2000)
9. Claypoole, R.L., Davis, G.M., Sweldens, W., Baraniuk, R.G.: Nonlinear wavelet transforms for image coding via lifting. IEEE Transactions on Image Processing 12, 1449–1459 (2003)
10. Habiboglu, Y.H., Kose, K., Cetin, A.E.: Fractional wavelet transform using an unbalanced lifting structure. In: Independent Component Analyses, Wavelets, Neural Networks, Biosystems, and Nanoengineering IX, Proc. SPIE, vol. 8058 (2011)
11. Pesquet-Popescu, B., Bottreau, V.: Three-dimensional lifting schemes for motion compensated video compression. In: IEEE International Conference on Acoustics, Speech, and Signal Processing, vol. 3, pp. 1793–1796 (2001)
12. Piella, G., Pesquet-Popescu, B., Heijmans, H.: Adaptive update lifting with a decision rule based on derivative filters. IEEE Signal Processing Letters 9, 329–332 (2002)
13. Heller, P.: Lagrange m-th band filters and the construction of smooth m-band wavelets. In: Proceedings of the IEEE-SP International Symposium on Time-Frequency and Time-Scale Analysis, 1994, pp. 108–111 (1994)

Visible and Infrared Image Registration Employing Line-Based Geometric Analysis

Jungong Han, Eric Pauwels, and Paul de Zeeuw

Centrum Wiskunde and Informatica (CWI)
Science Park 123, Amsterdam, The Netherlands

Abstract. We present a new method to register a pair of visible (ViS) and infrared (IR) images. Unlike most of existing systems that align interest points of two images, we align lines derived from edge pixels, because the interest points extracted from both images are not always identical, but most major edges detected from one image do appear in another image. To solve feature matching problem, we emphasize the geometric structure alignment of features (lines), instead of descriptor-based individual feature matching. This is due to the fact that image properties and patch statistics of corresponding features might be quite different, especially when one compares ViS image with long wave IR images (thermal information). However, the spatial layout of features for both images always preserves consistency. The last step of our algorithm is to compute the image transform matrix, given minimum 4 pairs of line correspondence. The comparative evaluation for algorithms demonstrates higher accuracy attained by our method when compared to the state-of-the-art approaches.[1]

Keywords: Image Registration, line detection, geometric analysis.

1 Introduction

Recent advances in imaging, networking, data processing and storage technology have resulted in an explosion in the use of multi-modality images in a variety of fields, including video surveillance, urban monitoring, cultural heritage area protection and many others. The integration of images from multiple channels can provide complementary information and therefore increase the accuracy of the overall decision making process. A fundamental problem in multi-modality image integration is that of aligning images of the same/similar scene taken by different modalities. This problem is known as image registration and the objective is to recover the correspondences between the images. Once such correspondences have been found, all images can be transformed into the same reference, enabling to augment the information in one image with the information from the others.

---

[1] This work is supported by EU-FP7 FIRESENSE project.

E. Salerno, A.E. Çetin, and O. Salvetti (Eds.): MUSCLE 2011, LNCS 7252, pp. 114–125, 2012.

1.1   Prior Work on Image Registration

Several related survey papers for image registration have appeared over the years. [1,2,3] have provided a broad overview of over three hundred papers for registering different types of sensors. Following most of literature, we also divide existing techniques into two categories: pixel-based methods and feature-based methods. Pixel-based methods first define a metric, such as the sum of square differences and mutual information [2], which measures the distance of two pixels from different images. The registration problem is then changed to minimize the total distance between all pixels on one image and the corresponding pixels on another image. In feature-based methods, interest points like Harris corners, scale invariant feature transform (SIFT), speed-up robust feature (SURF), etc., are first extracted from images. Afterwards, these features are matched based on the metrics, such as cross correlation and mutual information. Once more than four feature correspondences are obtained, the transform can be computed. In principle, pixel-based method should be better than the feature-based method because the former considers the global minimization of the cost function, but the later one minimizes the cost function locally. In practise, however, feature-based method has better performance for many applications, because the interest point is supposed to be distinctive in a local area, thus leading to the better matching. On the other hand, the pixel-based method is much more expensive than the feature-based algorithm, because every pixel is involved in the computation. Considering both accuracy and efficiency of the algorithm, we adopt the feature-based method in this paper. Therefore, we limit our review to feature-based registration methods, and pay special attention to the work for registering visible and infrared images.

Many approaches have been proposed for automatically registering IR and ViS images. Edge/gradient information is one of the most popular feature as their magnitudes [4] and orientations [5] may match between infrared and visible images. In [6], authors first extract edge segments, which are then grouped to form triangles. The transform can be computed by matching triangles from the source to destination images. Huang *et al.* [7] proposes a contour-based registration algorithm, which integrates the invariant moments with the orientation function of the contours to establish the correspondences of the contours in the two images. Normally it is difficult to obtain accurate registration by using contour-based method, because precisely matching all contours detected from two images is challenging. Moreover, this method drastically increases computation time compared to interest point-based registration. To improve this work, Han *et al.* [8] propose to find correspondences on *moving* contours. They extract silhouettes of moving humans from both images. Matching only the contours of humans significantly improves both the performance and the efficiency of the algorithm. An alternative [9] is to make use of the object moving pathes generated by object tracking algorithm. Finding correspondences between trajectories helps to align images. This type of algorithm works very well when moving objects can be precisely tracked from both channels. Unfortunately, the current tracking algorithm is not satisfactory in many applications.

Fig. 1. Interest point detection for IR and ViS images. Two different methods: SURF and Harris corner detection are used.

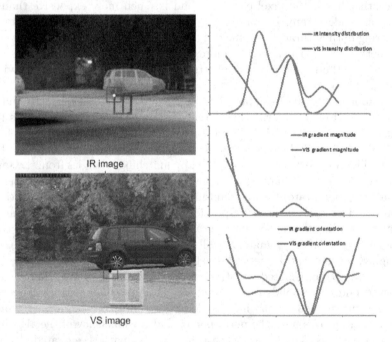

Fig. 2. Left: Statistics of corresponding points (within red square). Right, top: Distributions (normalized histogram) of intensity value. Middle: Distributions of gradient magnitude. Bottom: Distributions of gradient orientation. The x-axis represents the bin of histogram, and the y-axis refers to the number of pixels casted to the bin.

1.2 Problem Statement

Most existing publications for image registration dedicate to solving four problems: 1) an efficient way to extract feature points, which guarantees the majority of features on both images is identical; 2) a better feature descriptor; 3) a suitable metric to measure the distance of two feature descriptors; 4) a proper transform model. Among these four problems the detection of repeatable features and also the feature matching are more challenging when dealing with visual and infrared cameras. The main reason is that the electromagnetic wavelengths of ViS sensor and IR sensor are quite different. Normally, the wavelength of IR sensor is from 4 to 12 microns, while the wavelength of ViS sensor roughly lies between 0.4 to 0.7 microns. This leads to the fact that IR images have noticeably less texture in the area where temperatures are more homogeneous. However, the texture information is very important for both interest point detection and feature matching. In Fig. 1, we extract equivalent number of interest points from both IR and ViS images exploring two popular algorithms, where SURF method enables a scale- and rotation-invariant interest point detection but Harris method focuses on detecting corner points on the single scale. Seen from the results, the majority of extracted interest points is unfortunately not repeatable. To explain the feature matching problem, we show statistics (see Fig. 2) of image patches (15×15) surrounding two corresponding points. We compute the distribution (normalized histogram) of intensity value within the image patch, the distribution of gradient magnitude, and also the distribution of gradient orientation, respectively. Those are all feature descriptors widely used for IR and ViS image registration. To obtain a good feature matching result, we expect the signal distributions of two corresponding points to be similar to each other. Unfortunately, none of them is capable of measuring the correlation between two points in this case, though the gradient orientation is clearly better than the others. This example illustrates that comparing image patches may not be a reliable way to correlate features between IR and ViS images.

1.3 Our Contributions

In order to address two problems mentioned above, we propose a new algorithm here, which differentiates with existing work in two aspects. First, we do not use interest point as the feature to align the image. Instead, we try to align images based on lines derived from edges of the images. These lines strongly relate to the boundaries of objects, which always appear on both images though IR sensor and ViS sensor have significantly different properties. Secondly, our algorithm enables a one-to-many matching based on a simple feature descriptor, which allows one feature on one image to have more potential correspondences in another image. This ensures that the majority of the initial matching is correct. A central point of our feature-matching scheme is that it relies more on the geometric structure checking of features, which gives much better matching results. The last feature of our work is that we prove the traditional point-to-point transform can be directly computed, given minimum four pairs of line correspondence.

In the sequel, we first present our mathematical model in Section 2, which introduces how we compute the transformation matrix by employing lines. In Section 3, we describe several key algorithms, such as line reorganization, line initial matching and line-configuration computing. The experimental results are provided in Section 4. Finally, Section 5 draws conclusions and addresses our future research.

2   The Mathematical Model

The goal of image registration is to match two or more images so that identical coordinate points in these images correspond to the same physical region of the scene being imaged. To make our explanation simple, we assume that there are only two images involved in the registration, each representing one plane. In fact, the registration is to find a mathematical transformation model between these two images, which minimizes the *energy* function of image matching. This optimization procedure can be described mathematically

$$\tilde{\mathbf{H}} = arg \min_{\mathbf{H}} \sum_i E(p_i, \mathbf{H}p_i'). \tag{1}$$

Here, p_i is the ith pixel in the image I and p_i' is its corresponding pixel in the image I'. The energy function is to measure the *distance* between I and the transformed version of I' based on $\mathbf{H}$. This transformation helps to establish a plane-to-plane mapping, transforming a position p in one plane to the coordinate p' on another plane. The point $p = (u, v, w)^T$ in image coordinates corresponds to Euclidean coordinates $(u/w, v/w)$. In our paper, we assume a 2D perspective transformation. Writing positions as homogeneous coordinates, the transformation $p = \mathbf{H}p'$ equals

$$\begin{pmatrix} u \\ v \\ 1 \end{pmatrix} = \begin{pmatrix} h_{11} & h_{12} & h_{13} \\ h_{21} & h_{22} & h_{23} \\ h_{31} & h_{32} & h_{33} \end{pmatrix} \begin{pmatrix} u' \\ v' \\ 1 \end{pmatrix}. \tag{2}$$

Homogeneous coordinates are scaling invariant, reducing the degrees of freedom for the matrix H to only eight. In order to determine the eight parameters, at least four point-correspondences between two images have to be found. In the literature, most publications employ interest (corner) points for establishing point-correspondences. Normally, this matrix $\mathbf{H}$ is related to the camera model, that is, the matrix $\mathbf{H}$ can be further decomposed into camera intrinsic and extrinsic parameters.

As we mentioned before, our work wants to align images based on line correspondences. However, our objective is to compute a point-to-point transform matrix $\mathbf{H}$. Therefore, the central issues are whether and how we can obtain $\mathbf{H}$ based on line correspondences.

Let us now denote two lines (l and l') on both image coordinates as:

$$au + bv + c = 0 \quad and \quad a'u' + b'v' + c' = 0. \tag{3}$$

We can rewrite the line equation to

$$(a, b, c) \begin{pmatrix} u \\ v \\ 1 \end{pmatrix} = 0 \quad and \quad (a', b', c') \begin{pmatrix} u' \\ v' \\ 1 \end{pmatrix} = 0. \tag{4}$$

If we left multiply (a, b, c) to both sides of eqn. (2), it will become

$$(a, b, c) \begin{pmatrix} u \\ v \\ 1 \end{pmatrix} = (a, b, c)\mathbf{H} \begin{pmatrix} u' \\ v' \\ 1 \end{pmatrix} = 0. \tag{5}$$

Comparing (4) and (5), and taking into account that the line coefficients (a', b', c') for a given line are essentially unique (up to an arbitrary scaling factor), we can deduce that $(a, b, c)\mathbf{H} = \lambda(a', b', c')$. Here, λ is a scaling factor. If we write it in a formal way, we will obtain:

$$\mathbf{AH} = \mathbf{\Lambda A'}, \tag{6}$$

where $\mathbf{A}$ is a matrix, encoding the parameters of lines on one image coordinates, while $\mathbf{A'}$ is its corresponding matrix on another image coordinates. And, $\mathbf{\Lambda}$ encodes scaling factors for all lines. (5) turns out that it is possible to compute $\mathbf{H}$ directly from lines, given a number of pairs of corresponding lines. Next, we need to know how we compute the parameters of $\mathbf{H}$. Suppose that we have lines $l : (a, b, c)$ and $l' : (a', b', c')$ from different image coordinates, which correspond to each other. The associated scaling factor is λ_1. Therefore, we can get three equations, which are

$$\begin{aligned} ah_{11} + bh_{21} + ch_{31} &= \lambda_1 a' \\ ah_{12} + bh_{22} + ch_{32} &= \lambda_1 b' \\ ah_{13} + bh_{23} + ch_{33} &= \lambda_1 c'. \end{aligned} \tag{7}$$

If we divide the first equation by the third equation, and divide the second equation also by the third equation, we can remove the parameter λ, thereby achieving two linear equations:

$$\begin{aligned} ac'h_{11} + 0h_{12} - aa'h_{13} + bc'h_{21} + 0h_{22} - ba'h_{23} + cc'h_{31} + 0h_{32} &= ca'h_{33} \\ 0h_{11} + ac'h_{12} - ab'h_{13} + 0h_{21} + bc'h_{22} - bb'h_{23} + 0h_{31} + cc'h_{32} &= cb'h_{33}. \end{aligned} \tag{8}$$

Normally, we force h_{33} to 1. Therefore, we have 8 parameters to compute, and each pair of lines provides two equations related to $\mathbf{H}$. To have a complete matrix, at least four pairs of lines are required. The above deductions prove that it is possible to compute a point-to-point transform based on line correspondences.

3   Algorithm Implementation

In this section we will introduce algorithm implementations of two key modules in more details, which are line generation and line matching. The line generation module consists of line detection, line duplication deletion, line label and line sort. The line matching module includes initial matching and geometric matching of line composition. All steps are designed with the goal of constructing an efficient system.

3.1   Line Generation

RANSAC-based Hough transform from our previous work [10] is used to detect lines in the image. The output of our previous work is the start point and also the end point of a line. More precisely, it returns a line *segment*. The first step of our algorithm is to filter out some shorter line segments. For the rest of line segments, we will extend this segment until it goes through the entire image space, thereby leading to a *real* line. The reason for the first step is that line segments extracted from both images vary dramatically, but most *major* segments with sufficient length are identical on both images.

The Hough transform has the disadvantage that thick lines in the input image usually result in a bundle of detected lines, which all lie close together. In practice, we do not need so many lines, which are similar and close to each other. We expect to have only one representative line within certain area. To solve this problem, we introduce a line duplication deletion step after the Hough transform. Let a line obtained from the Hough transform be parameterized by its normal $\mathbf{n} = (n_x, n_y)^T$ with $\|\mathbf{n}\| = 1$ and the distance to the origin d. Two lines l_1, l_2 are considered equal if the angle between both is small, such as $\mathbf{n_1}^T \mathbf{n_2} > \cos(1.5°)$), and their distance is also small ($|d_1 - d_2| < 3$). The whole duplicate deletion process is repeated until the number of lines remains stable, which is usually after only three iterations.

Next, lines are labeled as either horizontal line or vertical line by

$$L_{hv} = \begin{cases} 1 & \text{if } |x_{end} - x_{start}| \geq |y_{end} - y_{start}|, \\ 0 & \text{otherwise}, \end{cases} \tag{9}$$

where x_{start}, y_{start} and x_{end}, y_{end} refer to x and y coordinates of start point and end point of a line, respectively. After labeling lines, the set of vertical lines are ordered left to right, the set of horizontal lines top to bottom. Later, when we will search for correspondences between images, we will put the constraint on the assignment that the order must be preserved. This constraint is likely valid in case that our transform is either affine transform or perspective transform.

Finally, the line is modeled by three parameters, which are L_{hv}, sp and os. If a line is labeled as a horizontal line ($L_{hv} = 1$), sp is defined as the angle of the line to the x−axis, and os means the offset of the line on the y−axis. The definitions for sp and os are just inverse if line is a vertical line. Fig. 3 shows the results after each step mentioned above. For this case, the number of lines detected by Hough transform is 20, but it reduces to 9 after processing.

3.2   Line Matching

As we can see from the problem statement part, feature initial matching schemes used by existing systems are in general not accurate enough. The main reason is that two images captured by different modalities are quite different at the pixel level. To solve this problem, our system enables a sort of one-to-many feature matching, which allows a line in one image to have several corresponding lines

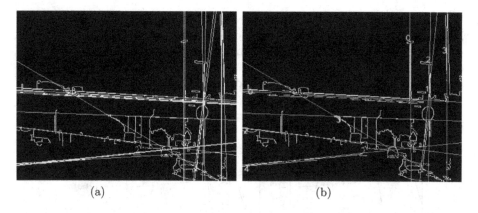

(a)   (b)

Fig. 3. Line generation. (a) Lines detected by Hough transform. (b) Lines after duplication deletion, HV labeling and HV sort. Horizontal lines and vertical lines are marked with different colors, and the number indicates the order of HV lines.

on another image. By doing so, we can ensure that several matching candidates must include the correct one. The basic idea for this initial matching is to check and compare three parameters of two lines located in two images. The first parameter is L_{hv}. We assume that two corresponding lines should have same label, which means the horizontal/vertical line in one image should correspond to a horizontal/vertical line on another image. The assumption is valid for most applications, where modalities are mounted on the same platform. The second parameter is sp, where we assume that corresponding lines have similar slope to the axis. The last parameter is used to compare distributions of the edge pixel surrounding the line. The surrounding area is the zone between two border lines, which have the same slope with the candidate line but with $\pm \epsilon$ offset shift, respectively. The distribution of the edge pixel within this area can be simply specified by the edge pixel percentage pec_{edge} of that area, equaling to N_{edge}/N_{total}. Here, N_{edge} refers to the number of edge pixels within that area, while N_{total} means the total number of pixels within that area. If we denote the parameters of two candidate lines as (L_{hv}, sp, pec_{edge}) and $(\tilde{L}_{hv}, \tilde{sp}, \tilde{pec}_{edge})$, our matching score S can thus be formulated as:

$$S = (L_{hv} == \tilde{L}_{hv}) \cdot K(\frac{sp - \tilde{sp}}{\sigma_{sp}}) \cdot K(\frac{pec_{edge} - \tilde{pec}_{edge}}{\sigma_{pec}}), \qquad (10)$$

where the first term of the left side $(L_{hv} == \tilde{L}_{hv})$ returns *true* if they are the same; otherwise it returns *false*. The rest two terms follow the same manner, in which $K(\cdot)$ is the Epanechnikov kernel function and σ indicates the width of the kernel, which can be set manually. The kernel function is specified by

$$K(y) = \begin{cases} 1 - |y|^2 & \text{for } |y|^2 \leq 1, \\ 0 & \text{otherwise.} \end{cases} \qquad (11)$$

We compute the matching scores between a given line and all candidate lines. Instead of selecting the best one, we allow one line to have multiple correspondences. The criterion is that we keep only one candidate if the matching score of this candidate is much higher than others. With the similar idea, we can adaptively assign up to three correspondences to a line.

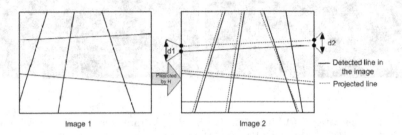

Fig. 4. An example for best line-configuration computing

After the step of line initial matching, we will process the geometric structure alignment of line compositions. The reason is that the matching between individual lines is not reliable due to the significant statistics difference between two image signals. However, the geometric structure (layout) of lines always remains consistency between two images. This observation motivates us to align the images by measuring the distance between two geometric structures formed by lines. The basic idea is that we randomly choose four lines from the first image to form a mini-configuration. Depending on the initial matching result, we will have several corresponding mini-configurations in the second image. This configuration-correspondence allows to compute the parameters of our eight-parameters perspective transform by solving a linear equation system according to formulas in Section 2. Using the obtained geometry transformation, we project one image onto the other image. The match between two images is evaluated by counting the total distance of the line to its closest projected line. We search for the transform parameters that provide the shortest distance by iterating over all configurations. This idea can be explained by a illustrated figure (Fig. 4), on which we transform the *image* 1 to the *image* 2. The black solid line represents the detected line on the image, and five red dash lines indicate the projected lines of the *image* 1 onto the *image* 2. From the mathematical point of view, finding the best configuration match equals to minimize a matching error M_e

$$M_e = \sum_{l \in \Phi} min(\|l', \mathbf{H}l\|_2, e_m), \tag{12}$$

where Φ the collection of lines in the *image* 1 and l' is the closest line of the projected line $\mathbf{H}l$ in the *image* 2. The metric $\|, \|_2$ denotes the Euclidean distance between the two lines, and the error for a line is bounded by a maximum value e_m.

The distance between two lines can be computed by summing up the distance of two *start* points d_1 and the distance of two *end* points d_2, which are illustrated in Fig. 4. Note that the start point and end point refer to the start and end point of a line on the image.

4   Experimental Results

We have tested our algorithm with 6 pairs of IR and ViS images, where 4 of them are outdoor scenarios[2] and 2 of them are describing indoor scenarios[3]. We show original images of both IR and ViS channels in Fig. 5, where the last three pairs are more challenging in terms of the focal length difference of two cameras.

We have registered these images by using our algorithm. A key parameter of our algorithm is the minimum length of the accepted line, for which we set 40 pixels. In general, our algorithm can register all pairs of images except the last one. The failure is caused by the fact that we cannot extract sufficient lines for geometric matching. To evaluate our registration algorithm, we measure and report the transform errors in Table 1. The transform error is measured by the distance between one point and its transformed corresponding point. More specifically, we randomly choose 5 salient points on IR image, and transform these 5 points to ViS image by using computed transform model. We manually label the corresponding points of those 5 points. The distance between the labeled point and the transformed point is proportional to the transform error.

Table 1. The measurement for transform errors

	pair 1	pair 2	pair 3	pair 4	pair 5
transform error	1.8 pixels	7.8 pixels	2.2 pixels	3.6 pixels	16.2 pixels

We also compared our line-based algorithm with algorithms based on interest point matching. Since gradient magnitude [4] and orientation [5] are widely used for IR and ViS image registration, our implementation explores statistics of gradient magnitude and orientation to describe the feature point, respectively. Afterwards, nearest neighbor approach is applied for feature matching. Next, RANSAC is used for rejecting some outliers. Finally, perspective transform matrix is computed based on a number of point correspondences between two images. We have tested these two feature descriptors for the same dataset. The gradient magnitude-based descriptor failed for all the pairs, and gradient orientation-based descriptor only succeeded in registering pair 2. We show the warped images in Fig. 6, where we warp the IR image to ViS image based on computed transform matrix. The results reveal that the registrations for pair 1, pair 3 and pair 4 are accurate. The registration for pair 2 is accepted for most

---

[2] Videos and images are provided by XenICs NV (Belgium).

[3] Images can be downloaded via
   http://www.dgp.toronto.edu/~nmorris/data/IRData/

Pair 1   Pair 2   Pair 3

Pair 4   Pair 5   Pair 6

Fig. 5. Original images for the experiment

Fig. 6. Warped images. The last one is generated by using gradient orientation-based feature matching.

parts, except for the right-upper corner of the image. The result for pair 5 is not good, but it is encouraging in the sense that two images are significantly different. In this figure, we also show the warped image of pair 2 by using the statistic of the gradient orientation as the feature descriptor.

5   Conclusion

In this paper, we have examined the use of line-correspondence for registering IR (long wavelength) and ViS images. Comparing with the interest point, line derived from edge pixels well represents the boundary of the object, which are always repeatable on images captured by different modalities. The feature matching module of our new method relies more on aligning the geometric structure of features, rather than matching individual feature only. Our new algorithm provides significant advantages over state-of-the-art approaches. The future work is the combination of global transform model used by this paper and the local transform model in order to further refine the registration locally.

References

1. Brown, L.: A Survey of Image Registration Techniques. ACM Computing Surveys 24(4), 325–376 (1992)
2. Zitova, B., Flusser, J.: Image Registration Methods: A Survey. Image and Vision Computing 21, 977–1000 (2003)
3. Xiong, Z., Zhang, Y.: A Critical Review of Image Registration Methods. Int. J. Image and Data Fusion 1(2), 137–158 (2010)
4. Lee, J., Kim, Y., Lee, D., Kang, D., Ra, J.: Robust CCD and IR Image Registration Using Gradient-Based Statistical Information. IEEE Signal Processing Letter 17(4), 347–350 (2010)
5. Kim, Y., Lee, J., Ra, J.: Multi-Sensor Image Registration Based on Intensity and Edge Orientation information. Pattern Recognition 41, 3356–3365 (2008)
6. Coiras, E., Santamaria, J., Miravet, C.: Segment-Based Registration Technique for Visual-Infrared Images. Optical Engineering 39, 282–289 (2000)
7. Huang, X., Chen, Z.: A Wavelet-Based Multisensor Image Registration Algorithm. In: Proc. ICSP, pp. 773–776 (2002)
8. Han, J., Bhanu, B.: Fusion of Color and Infrared Video for Moving Human Detection. Pattern Recognition 40, 1771–1784 (2007)
9. Caspi, Y., Simakov, D., Irani, M.: Feature-Based Sequence to Sequence Matching. Int. J. Comput. Vision 68(1), 53–64 (2006)
10. Han, J., Farin, D., de With, P.: Broadcast Court-Net Sports Video Analysis Using Fast 3-D Camera Modeling. IEEE Trans. Circuits Syst. Video Techn. 18(11), 1628–1638 (2008)

Texture Recognition Using Robust Markovian Features

Pavel Vácha and Michal Haindl

Institute of Information Theory and Automation of the ASCR,
Pod Vodarenskou Vezi 4, 182 08 Prague, Czech Republic
{vacha,haindl}@utia.cas.cz

Abstract. We provide a thorough experimental evaluation of several
state-of-the-art textural features on four representative and extensive
image databases. Each of the experimental textural databases ALOT,
Bonn BTF, UEA Uncalibrated, and KTH-TIPS2 aims at specific part of
realistic acquisition conditions of surface materials represented as multispectral textures. The extensive experimental evaluation proves the
outstanding reliable and robust performance of efficient Markovian textural features analytically derived from a wide-sense Markov random field
causal model. These features systematically outperform leading Gabor,
Opponent Gabor, LBP, and LBP-HF alternatives. Moreover, they even
allow successful recognition of arbitrary illuminated samples using a single training image per material. Our features are successfully applied
also for the recent most advanced textural representation in the form of
7-dimensional Bidirectional Texture Function (BTF).

Keywords: texture recognition, illumination invariance, Markov random fields, Bidirectional Texture Function, textural databases.

1 Introduction

Recognition of natural surface materials from their optical measurements represented as image textures, together with image (texture) segmentation, are the
inherent part of plethora of computer vision algorithms, which are exploited in
numerous real world applications such as visual scene analysis, image retrieval,
medical images segmentation, image compression, etc. The key issue in solving
real applications is robustness of employed methods, since images are usually
captured in real non-laboratory environment, where acquisition conditions such
as illumination, camera position, or noise cannot be controlled.

In this paper we focus on robustness of textural features to variations of illuminations conditions, such as spectrum, direction, and inhomogeneity. Illustrative
examples of such appearance variations are displayed in Figs. 3, 4, and 5. Possible theoretical approach to robust recognition is learning from images captured
under a full variety of possible illuminations for each material class [20,16], but it
is obviously impractical, expensive to acquire and compute, or even impossible,
if all needed measurements are not available. Alternatively, a kind of normalisation can be applied, e.g. cast shadow removal [7] or [8], which, unfortunately,

E. Salerno, A.E. Çetin, and O. Salvetti (Eds.): MUSCLE 2011, LNCS 7252, pp. 126–137, 2012.

completely wipes out rough texture structures with all their valuable discriminative information. Finally, the last and widely used approach is to construct corresponding invariants, which are features that do not change under specific variations of circumstances. However, it is necessary to keep in mind that an overdone invariance to broad range of sensing conditions inevitably reduces discriminability of features.

One of popular textural features are Local Binary Patterns [14] (LBP), which are invariant to any monotonic changes of pixel values, but they are very sensitive to noise [18] and illumination direction [19]. The LBP-HF extension [1] studies also relations between rotated patterns. Noise vulnerability was recently addressed by Weber Local Descriptor [4] (WLD). Texture similarity under different illumination direction [5] require the knowledge of illumination direction for all involved (trained as well as tested) textures. Finally, the MR8 texton representation [20] was extended to be colour and illumination invariant [2].

Multispectral textures can be described either jointly by multispectral textural features or separately by monospectral features on intensity image and colour features without spatial relations (histograms). The separate representation was advocated by [11], but we oppose this since a separate representation is not able to distinguish textures differing in position of pixels which have the same luminance. Obviously, the colour invariants computed from joint textural representation utilize the whole available information and they can create robust and compact texture description.

The contribution of this paper is a thorough evaluation of leading textural features under varying illumination spectrum, direction and slight variation of a camera location. We also test robustness to different acquisition devices, which is relevant especially for content-based image retrieval. These extensive tests of state of the art features were performed on four textural databases differing in variation of acquisition conditions and the results confirmed outstanding performance of Markovian textural features, preliminary tested in [19].

2   Markovian Textural Features

Our texture analysis is based on spatial and multimodal relations modelling by a wide-sense Markovian model. We employ a Causal Autoregressive Random (CAR) model, because it allows very efficient analytical estimation of its parameters. Subsequently, the estimated model parameters are transformed into illumination / colour invariants, which characterize the texture. These colour invariants encompass inter-spectral and spatial relations in the texture which are bounded to a selected contextual neighbourhood, see Fig. 1.

Let us assume that multispectral texture image is composed of C spectral planes (usually $C = 3$). $Y_r = [Y_{r,1}, \ldots, Y_{r,C}]^T$ is the multispectral pixel at location r, where the multiindex $r = [r_1, r_2]$ is composed of r_1 row and r_2 column index, respectively. The spectral planes are either modelled by 3-dimensional CAR model or mutually decorrelated by the Karhunen-Loeve transformation (Principal Component Analysis) and subsequently modelled using a set of C 2-dimensional CAR models.

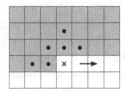 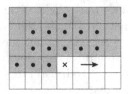

Fig. 1. Examples of contextual neighbourhood I_r. From the left, it is the unilateral hierarchical neighbourhood of third and sixth order. X marks the current pixel, the bullets are pixels in the neighbourhood, the arrow shows movement direction, and the grey area indicates acceptable neighbourhood pixels.

The CAR representation assumes that the multispectral texture pixel Y_r can be modelled as a linear combination of its neighbours:

$$Y_r = \gamma Z_r + \epsilon_r \ , \qquad Z_r = [Y_{r-s}^T : \forall s \in I_r]^T \tag{1}$$

where Z_r is the $C\eta \times 1$ data vector with multiindices r, s, t, $\gamma = [A_1, \ldots, A_\eta]$ is the $C \times C\eta$ unknown parameter matrix with square submatrices A_s . Some selected contextual causal or unilateral neighbour index shift set is denoted I_r and $\eta = cardinality(I_r)$, see Fig. 1. The white noise vector ϵ_r has normal density with zero mean and unknown full covariance matrix, same for each pixel.

The texture is analysed in a chosen direction, where multiindex t changes according to the movement on the image lattice. Given the known history of CAR process $Y^{(t-1)} = \{Y_{t-1}, Y_{t-2}, \ldots, Y_1, Z_t, Z_{t-1}, \ldots, Z_1\}$ the parameter estimation $\hat{\gamma}$ can be accomplished using fast and numerically robust statistics [9]:

$$\hat{\gamma}_{t-1}^T = V_{zz(t-1)}^{-1} V_{zy(t-1)} \ ,$$
$$V_{t-1} = \begin{pmatrix} \sum_{u=1}^{t-1} Y_u Y_u^T & \sum_{u=1}^{t-1} Y_u Z_u^T \\ \sum_{u=1}^{t-1} Z_u Y_u^T & \sum_{u=1}^{t-1} Z_u Z_u^T \end{pmatrix} + V_0 = \begin{pmatrix} V_{yy(t-1)} & V_{zy(t-1)}^T \\ V_{zy(t-1)} & V_{zz(t-1)} \end{pmatrix} \ , \tag{2}$$
$$\lambda_{t-1} = V_{yy(t-1)} - V_{zy(t-1)}^T V_{zz(t-1)}^{-1} V_{zy(t-1)} \ ,$$

where the positive definite matrix V_0 represents prior knowledge.

In the case of 2D CAR models stacked into the model equation (1), the uncorrelated noise vector components ϵ_r are additionally assumed. Consequently, the image spectral planes have to be decorrelated before modelling and the parameter matrices A_s are diagonal (in contrast with full matrices for general 3D CAR model).

Colour Invariants

Colour invariants are computed from the CAR parameter estimates to make them independent on changes of illumination intensity and colours. More precisely, these invariants are invariant to any linear change of pixel value vectors BY_r, where B is $C \times C$ regular transformation matrix. This is in accordance with reflectance models including specular reflections and even with the majority of

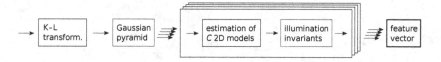

Fig. 2. Texture analysis algorithm using a set of 2D random field models

available BTFs [17], if illumination position remains unchanged. Additionally, 2D CAR models assume that the matrix B is diagonal. Moreover, our invariants are approximately invariant to infrequent changes of local illumination intensity and experiments show their robustness to variation of illumination direction. The following colour invariants were derived [18,17]:

1. trace: $\operatorname{tr} A_s$, $\forall s \in I_r$,
2. eigenvalues: $\nu_s = \operatorname{eigs}(A_s)$, $\forall s \in I_r$,
3. α_1: $1 + Z_r^T V_{zz}^{-1} Z_r$,
4. α_2: $\sqrt{\sum_r (Y_r - \hat{\gamma} Z_r)^T \lambda^{-1} (Y_r - \hat{\gamma} Z_r)}$,
5. α_3: $\sqrt{\sum_r (Y_r - \mu)^T \lambda^{-1} (Y_r - \mu)}$, μ is the mean value of vector Y_r .

The model parameters $\hat{\gamma}, \lambda$ are estimated using formula (2), we omit subsctripts for simplicity. Feature vectors are formed from these illumination invariants, which are easily evaluated during the CAR parameters estimation process.

In the case of 2D models, no eigenvalues are computed because matrices A_s are diagonal, and the features are formed from the diagonals without their re-ordering:

2. diagonals: $\nu_s = \operatorname{diag} A_s$, $\forall s \in I_r$.

Moreover, the invariants $\alpha_1 - \alpha_3$ are computed for each spectral plane separately.

Algorithm

The texture analysis algorithm starts with factorisation of texture image into K levels of the Gaussian down-sampled pyramid and subsequently each pyramid level is modelled by the CAR model. The pyramidal factorization is used, because it enables model to easily capture larger spatial relations. We usually use $K = 4$ levels of Gaussian down-sampled pyramid and the CAR models with the 6-th order semi-hierarchical neighbourhood (cardinality $\eta = 14$). If the image size is large enough (at least 400×400) it is possible to improve performance with the additional pyramid level $(K = 5)$. Finally, the estimated parameters for all pyramid levels are transformed into the colour invariants and concatenated into a common feature vector. The algorithm scheme for 2D CAR-KL is depicted in Fig. 2, where "-KL" suffix denotes decorrelation by Karhunen-Loeve transformation.

Fig. 3. Images from the UEA database, the upper row displays images under different illumination, while the bottom row shows images from different acquisition devices

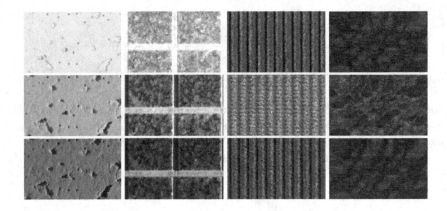

Fig. 4. Images from the Bonn BTF database, the first and second column with different illumination declination and the rest with various illumination azimuth

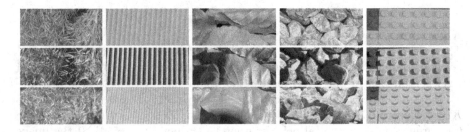

Fig. 5. Images from the ALOT database, each column shows images of the same material captured under varying illumination conditions

The dissimilarity between two feature vectors of two textures T, S is computed using fuzzy contrast [15] in its symmetrical form FC_3:

$$FC_3\,(T,S) = M - \left\{ \sum_{i=1}^{M} \min\left\{ \tau(f_i^{(T)}), \tau(f_i^{(S)}) \right\} - 3 \sum_{i=1}^{M} \left| \tau(f_i^{(T)}) - \tau(f_i^{(S)}) \right| \right\},$$

$$\tau(f_i) = \left(1 + \exp\left(-\frac{f_i - \mu(f_i)}{\sigma(f_i)} \right) \right)^{-1},$$

where M is the feature vector size and $\mu(f_i)$ and $\sigma(f_i)$ are average and standard deviation of the feature f_i computed over all database, respectively. The sigmoid function τ models the truth value of fuzzy predicate.

3   Experiments

We tested robustness of the CAR features in texture recognition on four different image data sets, each with different conditions. The first experiment is focused on recognition in variable illumination spectra and different acquisition devices, while the second experiment tests robustness to illumination direction changes. The next experiment utilises the largest recent collection of natural and artificial materials captured under various illumination conditions and the last test is classification into material categories. Summary of experiment setups is provided in Tab. 1.

Table 1. Parameters of experiments and comprised variations of recognition conditions

	Experiment			
texture database	UEA	Bonn	BTF ALOT	KTH
experiment conditions:				
illumination spectrum	+	–	+	–
illumination direction	–	+	+	–
viewpoint declination	–	–	+/–	–
acquisition device	+	–	–	–
experiment parameters:				
image size (bigger)	≈ 550	256	1536	200
number of classes	28	15,10	200, 250	11

The CAR features were compared with the most frequently used textural features as Gabor features [12], Opponent Gabor features [10], LBP [14], and LBP-HF [1]. These features demonstrated state of the art performance in the cited articles and all were tested with authors parameter settings. The grey level features such as Gabor features, LBP, and LBP-HF were computed either on grey level images or additionally for each spectral plane separately and concatenated, which is denoted with "RGB" suffix. For LBP features we tested variants

Table 2. Size of feature vectors

method	size	method	size
Gabor f.	144	2D CAR-KL	**260**
Opponent Gabor f.	**252**	3D CAR-KL	236
$LBP_{8,1+8,3,}$	512	2D CAR-KL (K=5)	325
$LBP_{8,1+8,3}$, RGB	**1536**	3D CAR-KL (K=5)	295
LBP-HF$_{8,1+16,2+24,3}$, RGB	1344	$LBP^{u2}_{16,2}$	243

$LBP_{8,1+8,3}$, $LBP^u_{16,2}$, and LBP-HF$_{8,1+16,2+24,3}$ reported by authors as the best in their experiments. Gabor features were additionally tested with and without separate normalisation of spectral planes (*Greyworld*), which is denoted with "norm." suffix. Size of feature vectors is summarised in Tab. 2. The following result figures display only the best performing features in each kind for each experiment.

3.1   University of East Anglia Uncalibrated Image Database

The first experiment was performed on UEA Uncalibrated Image Database[1] [6]. This dataset contains 28 textile designs, captured with 6 different devices (4 colour cameras and 2 colour scanners), and images for cameras were illuminated with 3 different illumination spectra, which sums up to 394 images in total. No calibration was performed and image resolution is about 550×450 (± 100). Examples of images are shown in Fig. 3.

In this experiment, training images per each material were randomly selected and the remaining images were classified using the Nearest Neighbour (1-NN) classifier, the results were averaged over 10^3 of random selections of training images. As it is displayed in Fig. 6, the alternative textural features were surpassed for all tested numbers of training images per material. It is quite surprising that LBP features had difficulties in this experiment, since they are invariant to any monotonic change of pixel values, while CAR features assume linear relation. UEA images are supposed to include even non-linear relations of images caused by different processing in acquisition devices. The poor performance of LBP features may be due to similarity of certain characteristics in UEA images that the LBP features are not able to distinguish or due to slight scale variation of images. The large images allowed to compute CAR features on $K = 5$ pyramid levels, the results for 2D CAR with $K = 4$ went from 56.6 to 85.3 for 1 to 6 training images, which still outperformed alternatives by a large margin.

3.2   Bonn BTF Database

The second experiment was performed on the University of Bonn BTF database [13], which consists of fifteen BTF colour measurements. Ten of those (corduroy, impalla, proposte, pulli, wallpaper, wool, ceiling, walk way, floor tile, pink tile)

---

[1] http://www.uea.ac.uk/cmp/research/graphicsvisionspeech/colour/data-code/

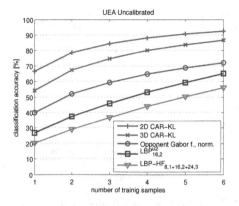
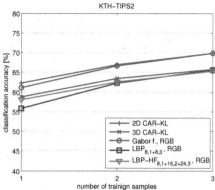

Fig. 6. Accuracy of classification on UEA images. The number of training images is changing from 1 to 6 per material.

Fig. 7. Accuracy of classification on KTH_TIPS2 database. The number of training material samples is changing from 1 to 3 per material category.

are now publicly available[2]. Each BTF material is measured in 81 illumination and 81 camera positions as an RGB image, examples of material appearance under varying illumination direction are shown in Fig. 4.

In our test set, we fixed viewpoint position to be perpendicular to material surface and included images under all 81 illumination positions. It is $15 \times 81 = 1215$ images in total, all were cropped to the same size 256×256 pixels. Training images per each material were again randomly selected and the remaining images were classified using 1-NN classifier. The number of training images went from 1 to 6 and the results were averaged over 10^3 of random selections of training images. The progress of classification accuracy is shown Fig. 8, where the CAR features outperformed the alternative features for all number of training images. The performance superiority of the CAR features is especially significant for low number of training samples, which confirms robustness of the CAR features to illumination direction variation. For additional details on robustness to illumination direction see [19,17].

3.3 Amsterdam Library of Textures

In this experiment, we tested the proposed features in the recognition of materials under combination of changing illumination spectrum and direction. The images of materials are from the recently created Amsterdam Library of Textures[3] (ALOT) [2]. The ALOT is a BTF database containing an extraordinary large collection of 250 materials, each acquired with varying viewpoint and illumination positions, and one additional illumination spectrum. Most of the materials have rough surfaces, so the movement of light source changes the appearance

---

[2] http://btf.cs.uni-bonn.de
[3] http://staff.science.uva.nl/~mark/ALOT/

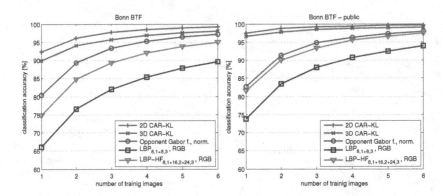

Fig. 8. Accuracy of classification on Bonn BTF database on the left and for its public part on the right. The number of training images is changing from 1 to 6 per material.

of materials. Moreover, the significant height variation of some materials (e.g. leaves) causes large and variable cast shadows, which make the recognition even more difficult.

In the "part a" of the experiment, we used one half of the dataset [2] with excluded multiple texture rotations. It contains images of the first 200 materials divided into training and test sets (1200 images each). Let c stands for camera, l for light, i for reddish illumination. The training set is defined as $c\{1,4\}l\{1,4,8\}$ and the test set contains setups $c\{2,3\}l\{3,5\}$, $c3l2$, and $c1i$. We cropped all the images to the same size 1536×660 pixels. The classification was evaluated on the test set images, where 1-NN classifier was trained on given numbers of images per material, all randomly selected from the training set.

In the "part b", we used images of all 250 materials, with all light setups, no rotations, and cameras 1 and 3, which is 14 images per material. Training images per material were randomly selected and the others were classified with 1-NN classifier, the results were averaged over 10^3 of random selections of training images. This test was performed separately for images from camera 1 and 3, the results were averaged (2×1750 images in total). As a consequence this part do not include recognition under viewpoint variation, which is in contrast with the "part a".

The results for both parts are displayed in Fig. 9, which shows the progress for different numbers of training images. The totally different scales of classification results are caused by images under different viewpoints included in the "part a" and the fact that none of the tested features are invariant to perspective projection. The viewpoint differences are even more extreme in the test set than in the training set. On the other hand, almost perfect results for 6 training images in the "part b" are not surprising, because 6 training images are leave-one-out methodology, which provides an upper bound on classification accuracy. In "part a", the CAR features outperformed the alternatives by 10% margin for all numbers of training images. In "part b", the performance of the CAR features is significantly better for low number of training images, while for leave-one-out

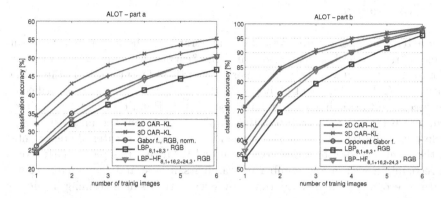

Fig. 9. Accuracy of classification on ALOT database, with the number of training images changing from 1 to 6 per material. It is worth to note different result scales caused by distinct difficulty of the setups.

the difference is about 1-2%. The *Greyworld* normalisation had minor effect on performance of Gabor and Opponent Gabor features.

The CAR features were computed on $K = 5$ pyramid levels, the results for 2D CAR with $K = 4$ went from 25.6 to 44.3 for "part a" and from 65.6 to 98.0 for "part b", which outperformed alternatives in "part b" and performed only slightly worse than Gabor features in "part a".

3.4 KTH-TIPS2 Database

Finally, the last experiment compares the performance of the proposed features on the KTH-TIPS2 database[4] [3], which includes material samples with different scales and rotations. However, as the training set always includes these scales and rotations, such an invariance is not an issue. The KTH-TIPS2 database contains 4 samples of 11 materials categories, each sample consists of images with 4 different illuminations, 3 in-plane rotations and 9 scales. The illumination conditions consist in 3 different directions plus 1 image with different spectrum. There are 4572 images and their resolution is varying around 200×200 pixels.

Training samples per each material category were again randomly selected and the remaining images were classified using 1-NN classifier. In this dataset, one training sample contains $4 \times 3 \times 9$ images, and we used from 1 to 3 samples per material category. Finally, the results were averaged over 10^3 of random selections of training samples.

Fig. 7 depicts the results, where all displayed features performed comparably. The reason is that each training sample includes images with all illumination variation, so any such invariance either do not matter or even may weaken discrimination of features. That is the reason, why Gabor features performed better without *Greyworld* normalisation.

---

[4] http://www.nada.kth.se/cvap/databases/kth-tips/

4   Conclusion

The extensive experimental evaluations illustrated in the paper prove the outstanding reliable and robust performance of efficient CAR illumination invariant Markovian textural features. The superiority of these features over leading alternatives as Gabor, Opponent Gabor, LBP, and LBP-HF features, was verified on the recent best available textural databases ALOT, Bonn BTF, UEA Uncalibrated, and KTH-TIPS2. These textural databases represent the majority of possible physically realistic acquisition conditions of surface materials represented in the form of visual textures. The proposed CAR features particularly excels in recognition with a low number of training samples, and they enable robust texture recognition in variable condition even with a single training image per material.

The results of the invariant texture retrieval or recognition can be checked online in our interactive demonstrations[5].

Acknowledgements. This research was supported by grant GAČR 102/08/0593 and partially by the projects MŠMT grants 1M0572 DAR, GAČR 103/11/0335, and CESNET 387/2010.

References

1. Ahonen, T., Matas, J., He, C., Pietikäinen, M.: Rotation Invariant Image Description with Local Binary Pattern Histogram Fourier Features. In: Salberg, A.-B., Hardeberg, J.Y., Jenssen, R. (eds.) SCIA 2009. LNCS, vol. 5575, pp. 61–70. Springer, Heidelberg (2009)
2. Burghouts, G.J., Geusebroek, J.M.: Material-specific adaptation of color invariant features. Pattern Recognition Letters 30, 306–313 (2009)
3. Caputo, B., Hayman, E., Mallikarjuna, P.: Class-specific material categorisation. In: Proceedings of the 10th IEEE International Conference on Computer Vision, ICCV 2005, October 17-21, pp. 1597–1604. IEEE (2005)
4. Chen, J., Shan, S., He, C., Zhao, G., Pietikäinen, M., Chen, X., Gao, W.: Wld: A robust local image descriptor. IEEE Transactions on Pattern Analysis and Machine Intelligence 32(9), 1705–1720 (2010)
5. Drbohlav, O., Chantler, M.: Illumination-invariant texture classification using single training images. In: Proceedings of the 4th International Workshop on Texture Analysis and Synthesis, Texture 2005, pp. 31–36 (2005)
6. Finlayson, G., Schaefer, G., Tian, G.: The UEA uncalibrated colour image database. Tech. Rep. SYS-C00-07, School of Information Systems, University of East Anglia, Norwich, United Kingdom (2000)
7. Finlayson, G., Hordley, S., Lu, C., Drew, M.: On the removal of shadows from images. IEEE Transactions on Pattern Analysis and Machine Intelligence 28(1), 59–68 (2006)
8. Finlyason, G., Xu, R.: Illuminant and gamma comprehensive normalisation in logRGB space. Patterm Recognition Letters 24, 1679–1690 (2002)

---

[5] http://cbir.utia.cas.cz

9. Haindl, M., Šimberová, S.: A Multispectral Image Line Reconstruction Method. In: Theory & Applications of Image Analysis, pp. 306–315. World Scientific Publishing Co., Singapore (1992)
10. Jain, A., Healey, G.: A multiscale representation including opponent colour features for texture recognition. IEEE Transactions on Image Processing 7(1), 124–128 (1998)
11. Mäenpää, T., Pietikäinen, M.: Classification with color and texture: jointly or separately? Pattern Recognition 37(8), 1629–1640 (2004)
12. Manjunath, B.S., Ma, W.Y.: Texture features for browsing and retrieval of image data. IEEE Transactions on Pattern Analysis and Machine Intelligence 18(8), 837–842 (1996)
13. Meseth, J., Müller, G., Klein, R.: Preserving realism in real-time rendering of bidirectional texture functions. In: OpenSG Symposium 2003, pp. 89–96. Eurographics Association, Switzerland (2003)
14. Ojala, T., Pietikäinen, M., Mäenpää, T.: Multiresolution gray-scale and rotation invariant texture classification with local binary patterns. IEEE Transactions on Pattern Analysis and Machine Intelligence 24(7), 971–987 (2002)
15. Santini, S., Jain, R.: Similarity measures. IEEE Transactions on Pattern Analysis and Machine Intelligence 21(9), 871–883 (1999)
16. Suen, P.H., Healey, G.: The analysis and recognition of real-world textures in three dimensions. IEEE Transactions on Pattern Analysis and Machine Intelligence 22(5), 491–503 (2000)
17. Vácha, P.: Query by pictorial example. Ph.D. thesis, Charles University in Prague, Prague (2011), http://cbir.utia.cas.cz/homepage/projects/phd_thesis/vacha_phd.pdf
18. Vacha, P., Haindl, M.: Image retrieval measures based on illumination invariant textural MRF features. In: Sebe, N., Worring, M. (eds.) Proceedings of ACM International Conference on Image and Video Retrieval, CIVR 2007, July 9-11, pp. 448–454. ACM (2007)
19. Vacha, P., Haindl, M.: Illumination invariants based on Markov random fields. In: Lovell, B., Laurendeau, D., Duin, R. (eds.) Proceedings of the 19th International Conference on Pattern Recognition, ICPR 2008, December 8-11, pp. 1–4. IEEE (2008)
20. Varma, M., Zisserman, A.: A statistical approach to texture classification from single images. International Journal of Computer Vision 62(1-2), 61–81 (2005)

A Plausible Texture Enlargement and Editing Compound Markovian Model

Michal Haindl and Vojtěch Havlíček

Institute of Information Theory and Automation
of the ASCR, Prague, Czech Republic
{haindl,havlicek}@utia.cz

Abstract. This paper describes high visual quality compound Markov random field texture model capable to realistically model multispectral bidirectional texture function, which is currently the most advanced representation of visual properties of surface materials. The presented compound Markov random field model combines a non-parametric control random field with analytically solvable wide-sense Markov representation for single regions and thus allows very efficient non-iterative parameters estimation as well as the compound random field synthesis. The compound Markov random field model is utilized for realistic texture compression, enlargement, and powerful automatic texture editing. Edited textures maintain their original layout but adopt anticipated local characteristics from one or several parent target textures.

Keywords: compound Markov random field, bidirectional texture function, texture editing, BTF texture enlargement.

1 Introduction

Physically correct virtual models with plausible appearance require not only complex 3D shapes accorded with the captured scene, but also object surfaces with realistic material appearances. Visual properties of such materials can be represented by advanced mathematical textural models capable to represent huge visual variability of real-world surface materials.

Because the appearance of real materials dramatically changes with illumination and viewing variations, any reliable representation of material visual properties requires capturing of its reflectance in as wide range of light and camera position combinations as possible. This is a principle of the recent most advanced texture representation, the Bidirectional Texture Function (BTF) [6]. The primary purpose of any synthetic texture approach is to reproduce and enlarge a given measured texture image so that ideally both natural and synthetic texture will be visually indiscernible. BTF function is represented by thousands of measurements (images) per material sample, thus its modelling prerequisite is simultaneously also significant compression capability[14].

Compound Markov random field models (CMRF) consist of several Markovian sub-models each having different characteristics along with an underlying

E. Salerno, A.E. Çetin, and O. Salvetti (Eds.): MUSCLE 2011, LNCS 7252, pp. 138–148, 2012.

structure model which controls transitions between these sub-models [18]. CMRF models were successfully applied to image restoration [7,18,21] or segmentation [27], however these models always require demanding numerical solutions with all their well known drawbacks.

Material appearance editing is useful approach which has large potential for significant speed-up and cost reduction in industrial virtual prototyping. A designer can propose several novel materials for objects in a virtual scene, checked them under anticipated illumination conditions, and select for production the best required option. Alternatively such editing process can simulate materials for which there are no available direct measurements. Unfortunately, image editing remains a complex user-directed task, often requiring proficiency in design, colour spaces, computer interaction and file management. Editing provides the scene designer with tools which enable to control virtual scene objects, geometric surfaces, illumination and objects faces appearance in the form of their corresponding textures. Image editing software is often characterised [5] by a seemingly endless array of toolbars, filters, transformations and layers. Although some recent attempts [1,2,20,17,25,24,3] have been made to automate this process, automatic integration of user preferences still remains an open problem in the context of texture editing [4,19]. Our fully automatic colour texture editing method [16] allows to synthesise and enlarge an artificial texture sharing anticipated frequency properties from both its parent textures.

We propose a non-interactive fully automatic texture editing approach based on our CMRF model [12] which combines a non-parametric and parametric analytically solvable MRFs and thus enable to avoid usual time consuming iterative Markov Chain Monte Carlo (MCMC) solution for both CMRF model parameters estimation as well as CMRF synthesis.

2   Compound Markov Model

Let us denote a multiindex $r = (r_1, r_2)$, $r \in {}^{\tau}I$, where ${}^{\tau}I$ is a discrete 2-dimensional rectangular lattice for the $\tau-th$ input random field, $\tau \in \{1, \ldots, N\}$, N (e.g. $N = 2$) is the number of materials represented by the corresponding CMRF models to be simultaneously combined, and r_1 is the row and r_2 the column index, respectively. Single 2-dimensional rectangular lattices ${}^{\tau}I$, can differ in their size. ${}^{\tau}X_r \in \{1, 2, \ldots, K_{\tau}\}$ is a random variable with natural number value (a positive integer), ${}^{\tau}Y_r$ is multispectral pixel at location r and ${}^{\tau}Y_{r,j} \in \mathcal{R}$ is its j-th spectral plane component. Both random fields $({}^{\tau}X, {}^{\tau}Y)$ are indexed on the same lattice ${}^{\tau}I$. Let us assume that each multispectral (BTF) observed texture ${}^{\tau}\tilde{Y}$ (composed of d spectral planes for all N CMRFs) can be modelled by a compound Markov random field model, where the principal discrete Markov random field (MRF) ${}^{\tau}X$ controls switching to a regional local MRF model ${}^{\tau}Y = \bigcup_{i=1}^{K_{\tau}} {}^{\tau i}Y$. Single K_{τ} regional submodels ${}^{\tau i}Y$ are defined on their corresponding lattice subsets ${}^{\tau i}I$, ${}^{\tau i}I \cap {}^{\tau j}I = \emptyset$ $\forall i \neq j$, ${}^{\tau}I = \bigcup_{i=1}^{K_{\tau}} {}^{\tau i}I$ and they are of the same MRF type. They differ only in their contextual support sets ${}^{\tau i}I_r$ and corresponding parameters sets ${}^{\tau i}\theta$. Note, that this condition

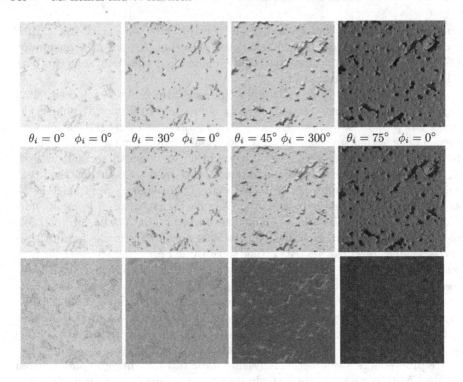

$\theta_i = 0° \quad \phi_i = 0°$ $\theta_i = 30° \quad \phi_i = 0°$ $\theta_i = 45° \quad \phi_i = 300°$ $\theta_i = 75° \quad \phi_i = 0°$

Fig. 1. BTF ceiling panel texture measurements (upper row), their synthetic counterparts (middle row), and edited measurements combined with wood material (bottom row) for various elevation (θ_i) and azimuthal (ϕ_i) illumination angles

is used only to simplify the resulting algorithm and can be easily removed if required. The parametric local MRF type should be preferably a MRF model with a fast non iterative syntheses, such as the Gaussian MRF defined on a toroidal lattice. The CMRF model has posterior probability

$$P(^\tau X, {}^\tau Y \mid {}^\tau \tilde{Y}) = P(^\tau Y \mid {}^\tau X, {}^\tau \tilde{Y}) \, P(^\tau X \mid {}^\tau \tilde{Y})$$

and the corresponding optimal MAP solution is:

$$(^\tau \hat{X}, {}^\tau \hat{Y}) = \arg \max_{^\tau X \in \Omega_{\tau X}, {}^\tau Y \in \Omega_{\tau Y}} P(^\tau Y \mid {}^\tau X, {}^\tau \tilde{Y}) \, P(^\tau X \mid {}^\tau \tilde{Y}) \ ,$$

where $\Omega_{\tau X}, \Omega_{\tau Y}$ are the corresponding configuration spaces for random fields $(^\tau X, {}^\tau Y)$. To avoid iterative MCMC MAP solution, we propose the following two step approximation for all CMRFs:

$$(^\tau \breve{X}) = \arg \max_{^\tau X \in \Omega_{\tau X}} P(^\tau X \mid {}^\tau \tilde{Y}) \ ,$$

$$(^\tau \breve{Y}) = \arg \max_{^\tau Y \in \Omega_{\tau Y}} P(^\tau Y \mid {}^\tau \breve{X}, {}^\tau \tilde{Y}) \ .$$

This approximation significantly simplifies CMRF estimation because it allows to take advantage of simple analytical estimation of regional parametric MRF models used in the presented model but is advantageous also for any other possible alternative parametric MRF model. The following two subsections hold for all N CMRFs thus their index τ is omitted to simplify notation.

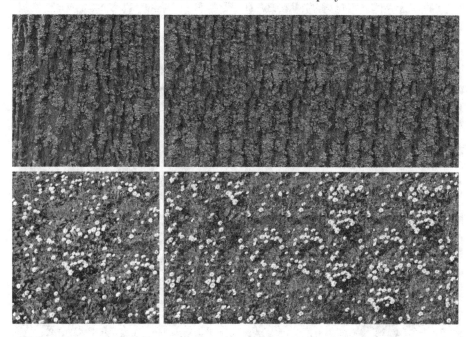

Fig. 2. Synthetic enlarged colour maple bark and grass textures estimated from their natural measurements (left column)

2.1   Region Switching Markov Model

The principal MRF $(P(X \mid \tilde{Y}))$ can be, for example, represented by a flexible K−state Potts random field [23,26]. Instead of this or some alternative parametric MRF, which require a MCMC solution, we suggest to use simple non-parametric approximation based on our roller method [9,10]. The roller was chosen because it is fully automatic and very fast method which produces high quality spatial data enlargement results (see details in [9,10]).

The control random field $\check{X}$ is estimated using simple K-means clustering of $\tilde{Y}$ in the RGB colour space into predefined number of K classes, where cluster indices are $\check{X}_r \quad \forall r \in I$ estimates. The number of classes K can be estimated using the Kullback-Leibler divergence and considering sufficient amount of data necessary to reliably estimate all local Markovian models.

The roller method is subsequently used for optimal $\check{X}$ compression and extremely fast enlargement to any required random field size. The roller method [9,10] is based on the overlapping tiling and subsequent minimum error boundary cut. One or several optimal double toroidal data patches are seamlessly repeated during the synthesis step. This fully automatic method starts with the minimal tile size detection which is limited by the size of control field, the number of toroidal tiles we are looking for and the sample spatial frequency content.

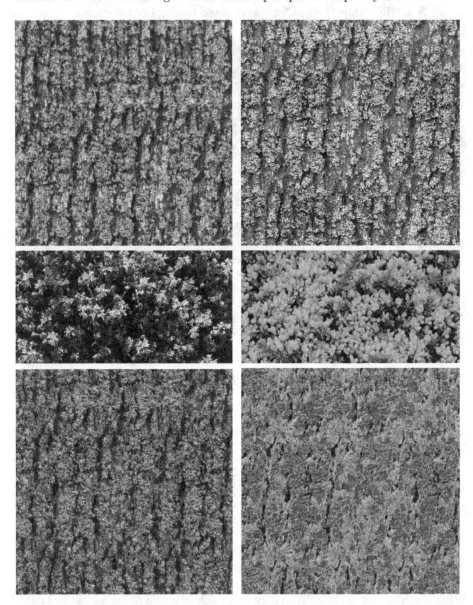

Fig. 3. Synthetic edited cutouts from the enlarged colour maple bark texture models (odd rows). The upper left result took over some local models from the grass texture on Fig.2, bottom textures learn some models on the corresponding even row real textures.

2.2   Local Markov Models

Local i-th texture region (not necessarily continuous) is represented by the adaptive 3D causal autoregressive random (3DCAR) field model [13,11] because this model can be analytically estimated as well as easily synthesised. Alternatively we could use a 3D Gaussian Markov random field (GMRF) or spectrally decorrelated 2D CAR or GMRF models [8,15]. All these models allow analytical synthesis (see [8] for the corresponding conditions) and they can be unified in the following matrix equation form (i-th model index is further omitted to simplify notation):

$$Y_r = \gamma \, Z_r + \epsilon_r \ , \tag{1}$$

where

$$Z_r = [Y_{r-s}^T : \forall s \in I_r]^T \tag{2}$$

is the $\eta d \times 1$ data vector with multiindices r, s, t, $\gamma = [A_1, \ldots, A_\eta]$ is the $d \times d\,\eta$ unknown parameter matrix with parametric submatrices A_s. In the case of d 2D CAR / GMRF models stacked into the model equation (1) the parameter matrices A_s are diagonal otherwise they are full matrices for general 3DCAR models [11]. The model functional contextual neighbour index shift set is denoted I_r and $\eta = cardinality(I_r)$. GMRF and CAR models mutually differ in the correlation structure of the driving noise ϵ_r (1) and in the topology of the contextual neighbourhood I_r (see [8] for details). As a consequence, all CAR model statistics can be efficiently estimated analytically [13] while the GMRF statistics estimates require either numerical evaluation or some approximation ([8]).

Given the known 3DCAR process history

$$Y^{(t-1)} = \{Y_{t-1}, Y_{t-2}, \ldots, Y_1, Z_t, Z_{t-1}, \ldots, Z_1\}$$

the parameter estimation $\hat{\gamma}$ can be accomplished using fast, numerically robust and recursive statistics [13]:

$$\hat{\gamma}_{t-1}^T = V_{zz(t-1)}^{-1} V_{zy(t-1)} \ ,$$

$$V_{t-1} = \tilde{V}_{t-1} + V_0 \ ,$$

$$\tilde{V}_{t-1} = \begin{pmatrix} \sum_{u=1}^{t-1} Y_u Y_u^T & \sum_{u=1}^{t-1} Y_u Z_u^T \\ \sum_{u=1}^{t-1} Z_u Y_u^T & \sum_{u=1}^{t-1} Z_u Z_u^T \end{pmatrix}$$

$$= \begin{pmatrix} \tilde{V}_{yy(t-1)} & \tilde{V}_{zy(t-1)}^T \\ \tilde{V}_{zy(t-1)} & \tilde{V}_{zz(t-1)} \end{pmatrix} \ ,$$

$$\lambda_{t-1} = V_{yy(t-1)} - V_{zy(t-1)}^T V_{zz(t-1)}^{-1} V_{zy(t-1)} \ ,$$

where V_0 is a positive definite matrix (see [13]). Although, an optimal causal (for (2D/3D)CAR models) functional contextual neighbourhood I_r can be solved analytically by a straightforward generalisation of the Bayesian estimate in [13], we use faster approximation which does not need to evaluate statistics for all possible I_r configurations. This approximation is based on large spatial correlations. We start from the causal part of a hierarchical non-causal neighbourhood and neighbours locations corresponding to spatial correlations larger than a specified threshold (> 0.6) are selected. The i-th model synthesis is simple direct application of (1) for both 2DCAR or 3DCAR models. A GMRF synthesis requires one FFT transformation at best [8]. 3DCAR models provide better spectral modelling quality than the alternative spectrally decorrelated 2D models for motley textures at the cost of small increase of number of parameters to be stored.

2.3   Texture Editing

An edited $\tau-th$ texture maintains its original layout i.e. the $^\tau X$ non-parametric control field and corresponding sublattices $^\tau I = \bigcup_{i=1}^{K_\tau} {}^{\tau i}I$ but adopt local characteristics from one or several parent target textures:

$$^\tau Y^\epsilon = \bigcup_{i=1}^{K_\tau} {}^{\pi_i i}Y \ ,$$

where $\pi_i \in \{1_1, \ldots, K_1, \ldots, N_1, \ldots, K_N\}$ and $\pi_i \leq K_\tau$. It is sufficient to replace only models corresponding to larger sublattices $^{\tau i}I$ to obtain interesting visual effects. Numerous editing strategies are possible. Model mapping can be based on similar size index sublattices, similar colours, similar local correlation structure, or local frequencies, and many others. Simultaneously it is possible to keep some original material models, for example to keep the bark texture model in Fig.2 but replace the lichen model for some alternative one.

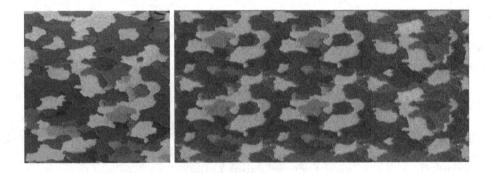

Fig. 4. Bark texture and its enlarged version

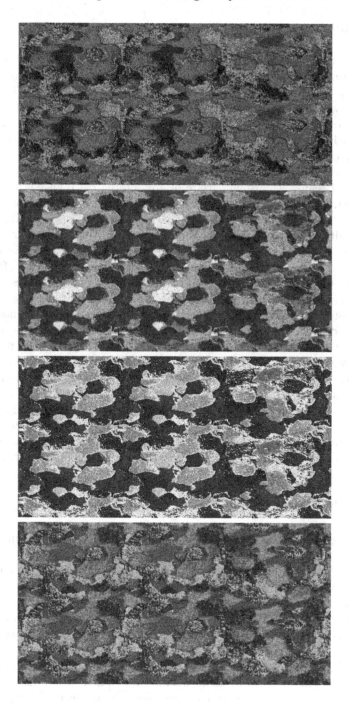

Fig. 5. Bark texture edited enlarged versions (seaweed, flower, pebbles, plant learning parent textures - downwards)

3   Results

We have tested the presented CMRF model-based editing approach on natural colour textures from our extensive texture database (http://mosaic.utia.cas.cz, Fig.2-bark), which currently contains over 1000 colour textures, CGTextures (http://www.cgtextures.com, Fig.2-grass), and on BTF measurements from the University of Bonn [22] (Fig.1-upper row). Tested edited textures were either natural, such as two textures on Fig.2 and bark texture on Fig.4, or man-made Fig.1 (ceiling panel). All edited results combined only two different surface materials ($N = 2$) and used the model mapping strategy based on similar size index sublattices. Because we are not aware of any alternative BTF editing method allowing partial exchange of local surface models, we could not compare our results with some alternative approach. Each BTF material sample included in the University of Bonn database [6] is measured in 81 illumination and viewing angles, respectively. A material sample measurements (Fig.1) from this database have resolution of 800×800 and size 1.2 GB. Fig.1-upper row shows four such measurements of ceiling panel material for different illumination angles and fixed perpendicular view (elevation and azimuthal view angles are zero $\theta_v = \phi_v = 0°$). Examples on Figs.1,2 use six level control field $(K = 6)$ and causal neighbourhood derived from the 20th order hierarchical contextual neighbourhood.

Resulting synthetic more complex textures (such as grass with flowers on Fig.2) have generally better visual quality (there is no any usable analytical quality measure) than textures synthesised by previously published [8,15,11,14] alternative simpler MRF models (Fig.2-bottom row). The BTF-CMRF variant of the model uses similar fundamental flowchart with our Markovian BTF model (see details in [14]) except the range map estimation, range map modelling, and displacement filter steps, respectively. BTF-CMRF is capable to reach huge BTF compression ration $\sim 1 : 1 \times 10^5$ relative to the original BTF measurements but $\approx 5\times$ lower than [14].

4   Conclusions

The presented CMRF (BTF-CMRF) modelling and editing method shows good performance for visual quality, compression, and numerical efficiency for the most of our tested real-world materials. The model allows huge compression ratio (only tens of parameters per BTF and few small control field tiles) and exceptionally fast seamless synthesis of any required texture size. The method does not allow unseen (unmeasured) BTF space data restoration or modelling, unlike some fully parametric probabilistic BTF models, due to the non-parametric model component.

The editing application of this model is fully automatic and creates anticipated visual results without need of usual manual interaction of alternative texture editing approaches, provided we can use some automatic model replacement strategy.

Acknowledgements. This research was supported by grant GAČR 102/08/0593 and partially by the MŠMT grant 1M0572 DAR, GAČR 103/11/0335, CESNET 387/2010.

References

1. Ashikhmin, M.: Synthesizing natural textures. Tech. rep. (2001)
2. Bar-Joseph, Z., El-Yaniv, R., Lischinski, D., Werman, M.: Texture mixing and texture movie synthesis using statistical learning. IEEE Transactions on Visualization and Computer Graphics 7(2), 120–135 (2001)
3. Brooks, S., Cardle, M., Dodgson, N.A.: Enhanced texture editing using self similarity. In: VVG, pp. 231–238 (2003)
4. Brooks, S., Dodgson, N.A.: Self-similarity based texture editing. ACM Trans. Graph. 21(3), 653–656 (2002)
5. Brooks, S., Dodgson, N.A.: Integrating procedural textures with replicated image editing. In: Spencer, S.N. (ed.) Proceedings of the 3rd International Conference on Computer Graphics and Interactive Techniques in Australasia and Southeast Asia 2005, Dunedin, New Zealand, November 29-December 2, pp. 277–280. ACM (2005)
6. Filip, J., Haindl, M.: Bidirectional texture function modeling: A state of the art survey. IEEE Transactions on Pattern Analysis and Machine Intelligence 31(11), 1921–1940 (2009)
7. Geman, S., Geman, D.: Stochastic relaxation, gibbs distributions and bayesian restoration of images. IEEE Trans. Pattern Anal. Mach. Int. 6(11), 721–741 (1984)
8. Haindl, M.: Texture synthesis. CWI Quarterly 4(4), 305–331 (1991)
9. Haindl, M., Hatka, M.: BTF Roller. In: Chantler, M., Drbohlav, O. (eds.) Proceedings of the 4th International Workshop on Texture Analysis, Texture 2005, pp. 89–94. IEEE, Los Alamitos (2005)
10. Haindl, M., Hatka, M.: A roller - fast sampling-based texture synthesis algorithm. In: Skala, V. (ed.) Proceedings of the 13th International Conference in Central Europe on Computer Graphics, Visualization and Computer Vision, pp. 93–96. UNION Agency - Science Press, Plzen (2005)
11. Haindl, M., Havlíček, V.: A multiscale colour texture model. In: Kasturi, R., Laurendeau, D., Suen, C. (eds.) Proceedings of the 16th International Conference on Pattern Recognition, pp. 255–258. IEEE Computer Society, Los Alamitos (2002), http://dx.doi.org/10.1109/ICPR.2002.1044676
12. Haindl, M., Havlíček, V.: A compound mrf texture model. In: Proceedings of the 20th International Conference on Pattern Recognition, ICPR 2010, pp. 1792–1795. IEEE Computer Society CPS, Los Alamitos (2010), http://doi.ieeecomputersociety.org/10.1109/ICPR.2010.442
13. Haindl, M., Šimberová, S.: A Multispectral Image Line Reconstruction Method. In: Theory & Applications of Image Analysis, pp. 306–315. World Scientific Publishing Co., Singapore (1992)
14. Haindl, M., Filip, J.: Extreme compression and modeling of bidirectional texture function. IEEE Transactions on Pattern Analysis and Machine Intelligence 29(10), 1859–1865 (2007)
15. Haindl, M., Havlíček, V.: A Multiresolution Causal Colour Texture Model. In: Amin, A., Pudil, P., Ferri, F., Iñesta, J.M. (eds.) SPR 2000 and SSPR 2000. LNCS, vol. 1876, pp. 114–122. Springer, Heidelberg (2000)

16. Haindl, M., Havlíček, V.: Texture Editing Using Frequency Swap Strategy. In: Jiang, X., Petkov, N. (eds.) CAIP 2009. LNCS, vol. 5702, pp. 1146–1153. Springer, Heidelberg (2009)
17. Hertzmann, A., Jacobs, C.E., Oliver, N., Curless, B., Salesin, D.H.: Image analogies. ACM Trans. Graph., 327–340 (2001)
18. Jeng, F.C., Woods, J.W.: Compound gauss-markov random fields for image estimation. IEEE Transactions on Signal Processing 39(3), 683–697 (1991)
19. Khan, E.A., Reinhard, E., Fleming, R.W., Bülthoff, H.H.: Image-based material editing. ACM Trans. Graph. 25(3), 654–663 (2006),
    http://doi.acm.org/10.1145/1141911.1141937
20. Liang, L., Liu, C., Xu, Y.Q., Guo, B., Shum, H.Y.: Real-time texture synthesis by patch-based sampling. ACM Transactions on Graphics (TOG) 20(3), 127–150 (2001)
21. Molina, R., Mateos, J., Katsaggelos, A., Vega, M.: Bayesian multichannel image restoration using compound gauss-markov random fields. IEEE Trans. Image Proc. 12(12), 1642–1654 (2003)
22. Müller, G., Meseth, J., Sattler, M., Sarlette, R., Klein, R.: Acquisition, synthesis and rendering of bidirectional texture functions. In: Eurographics 2004. STAR - State of The Art Report, pp. 69–94. Eurographics Association (2004)
23. Potts, R., Domb, C.: Some generalized order-disorder transformations. In: Proceedings of the Cambridge Philosophical Society, vol. 48, p. 106 (1952)
24. Wang, X., Wang, L., Liu, L., Hu, S., Guo, B.: Interactive modeling of tree bark. In: Proc. 11th Pacific Conf. on Computer Graphics and Applications, pp. 83–90. IEEE (2003)
25. Wiens, A.L., Ross, B.J.: Gentropy: evolving 2d textures. Computers & Graphics 26, 75–88 (2002)
26. Wu, F.: The Potts model. Reviews of modern physics 54(1), 235–268 (1982)
27. Wu, J., Chung, A.C.S.: A segmentation model using compound markov random fields based on a boundary model. IEEE Trans. Image Processing 16(1), 241–252 (2007)

Bidirectional Texture Function Simultaneous Autoregressive Model

Michal Haindl and Michal Havlíček

Institute of Information Theory and Automation
of the ASCR, Prague, Czech Republic
{haindl,havlimi2}@utia.cz

Abstract. The Bidirectional Texture Function (BTF) is the recent most advanced representation of visual properties of surface materials. It specifies their altering appearance due to varying illumination and viewing conditions. Corresponding huge BTF measurements require a mathematical representation allowing simultaneously extremal compression as well as high visual fidelity. We present a novel Markovian BTF model based on a set of underlying simultaneous autoregressive models (SAR). This complex but efficient BTF-SAR model combines several multispectral band limited spatial factors and range map sub-models to produce the required BTF texture space. The BTF-SAR model enables very high BTF space compression ratio, texture enlargement, and reconstruction of missing unmeasured parts of the BTF space.

Keywords: BTF, texture analysis, texture synthesis, data compression, virtual reality.

1 Introduction

Realistic virtual scenes requires object faces covered with synthetic textures visually as close as possible to the corresponding real surface materials they imitate. Such textures have to model rugged surfaces, do not obey Lambertian law, and their reflectance is illumination and view angle dependent. Their most advanced recent representation is the Bidirectional Texture Function (BTF) [3,7,17] which is a 7-dimensional function describing surface texture appearance variations due to varying illumination and viewing angles. Such a function is typically represented by thousands of images per material sample, each taken for a specific combination of the illumination and viewing condition. Visual textures can be either represented by digitised natural textures or textures synthesised from an appropriate mathematical model.

The former simplistic alternative suffers among others with extreme memory requirements for storage of a large number of digitised cross sectioned slices through different material samples or measured BTF space (apposite example can be found in [21]). Sampling solution become even unmanageable for correctly modeled visual scenes with BTF texture representation which require to store tens thousands of different illumination and view angle samples for every texture

E. Salerno, A.E. Çetin, and O. Salvetti (Eds.): MUSCLE 2011, LNCS 7252, pp. 149–159, 2012.

so that even simple visual scene with several materials requires to store tera bytes of texture data which is still far out of limits for any current and near-future hardware. Several intelligent sampling methods (for example [4,5] and many others) were proposed to reduce these extreme memory requirements. All these methods are based on some sort of original small texture sampling and the best of them produce very realistic synthetic textures. However they require to store thousands images for every combination of viewing and illumination angle of the original target texture sample and in addition often produce visible seams (except for method presented in [13]). Some of them are computationally demanding and they are not able to generate textures unseen [17] by these algorithms.

Contrary to the sampling approaches, the synthetic textures generated from mathematical models are more flexible and extremely compressed, because several parameters have to be stored only. They may be evaluated directly in a procedural form and can be used to fill virtually infinite texture space without visible discontinuities. On the other hand, mathematical models can only approximate real measurements, which results in visual quality compromise for some oversimplified methods. Several multispectral smooth modelling approaches were published - consult for example [9,15,1,18,16,12,11]. Modelling static BTF textures requires seven dimensional models, but it is possible to approximate this general BTF model with a set of much simpler three or two dimensional factorial models, provided we will accept some information loss.

Among such possible models the random fields are appropriate for texture modelling not only because they do not suffer from some problems of alternative options (see [9,18] for details), but they also provide easy texture synthesis and sufficient flexibility to reproduce a large set of both natural and artificial textures. While the random field based models quite successfully represent high frequencies present in natural textures, low frequencies are sometimes difficult for them. This slight drawback may be overcome by using a multiscale random field model. Multiple resolution decomposition such as Gaussian Laplacian pyramids, wavelet pyramids or subband pyramids present efficient method for the spatial information compressing. The hierarchy of different resolutions of an input image provides a transition between pixel level features and region or global features and hence such a representation simplify modelling a large variety of possible textures. Each resolution component is both analysed and synthesised independently.

We propose a novel algorithm for efficient rough texture modelling which combines an estimated range map with synthetic multiscale SAR based generated smooth texture. The texture visual appearance during changes of viewing and illumination conditions are simulated using the bump mapping [2] or displacement mapping [22] technique. The obvious advantage of this solution is the possibility to use hardware support for both bump and displacement mapping techniques in the contemporary visualisation hardware (GPU).

2   BTF-SAR Model

The BTF-SAR model combines an estimated range map with synthetic multi-scale smooth texture. The overall BTF-SAR model scheme is on Fig.1. The model starts with range map estimation (section 2.1) followed by the BTF illumination / view $(\theta_i, \phi_i / \theta_v, \phi_v)$ space segmentation into c subspace images (the closest BTF images to cluster centers) using the K-means algorithm. Eigen-analysis of BTF data has shown that $c = 20$ is sufficient to represent its reflectance correctly for most of the material samples. The color cumulative histograms of individual BTF images, in perceptually uniform CIE Lab color-space, are used as the data features. Smooth parts of single BTF subspace spatial factors (section 2.2) textures $(3D\ Y)$ are modelled using the SAR factorial texture model of section 3. Each multispectral fine-resolution BTF subspace component is obtained from the pyramid collapse procedure (i.e. the interpolation of sub-band components - the inversion process to the creation of the Gaussian-Laplacian pyramid).

The overall BTF texture visual appearance during changes of viewing and illumination conditions is simulated using either bump or displacement mapping technique. This solution can benefit from bump / displacement mapping hardware support in contemporary visualisation hardware.

Let us denote multiindices r, s $r = (r_1, r_2), r \in I$ or $r \in \tilde{I}$, (similarly for $s = (s_1, s_2)$) where $I, \tilde{I}$ are discrete 2-dimensional finite rectangular lattices with toroidal border conditions indexing the model random field and measured BTF images, respectively. Usually, the synthesized random field is much larger than the measured one, i.e. $\tilde{I} \subset I$. r_1 is the row, and r_2 the column index, respectively. Y_r is multispectral pixel at location r and $Y_{r,j} \in \mathcal{R}$ is its j-th spectral plane component $(j \in< 1; d >)$.

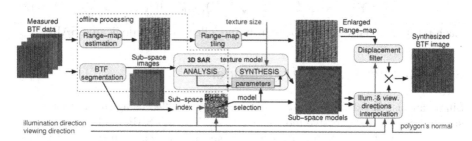

Fig. 1. The BTF-SAR modelling scheme

2.1   Range Map Modelling

The overall roughness of a textured surface significantly influences the BTF texture appearance. Such a surface can be specified using its single range map per material, which can be either measured or estimated by several existing approaches such as the shape from shading [8], shape from texture [6] or photometric stereo [23], respectively. The BTF-SAR range map estimate benefits from tens of ideally mutually registered BTF measurements (e.g. 81 for a fixed view of the University of Bonn data) and uses the over-determined photometric stereo from mutually aligned BTF images. The estimated material range map is further enlarged by the image roller technique [13,14]. The photometric stereo enables to acquire the normal and albedo fields from at least three intensity images obtained for different illuminations but fixed camera position while a Lambertian opaque surface is assumed. However, the photometric stereo method is not well suited for surfaces with highly specular reflectance, highly subsurface scattering or strong occlusion, since it breaks the underlying Lambertain reflectance assumption.

2.2 Spatial Factorization

An analysed BTF subspace texture is decomposed into multiple resolutions factors using the Laplacian pyramid and the intermediary Gaussian pyramid $Y_{\bullet}^{''(k)}$ which is a sequence of images in which each one is a low-pass down-sampled version of its predecessor. The symbol $\bullet$ denotes all corresponding indices $(\forall r \in I)$. The Gaussian pyramid for a reduction factor n is [11]:

$$Y_r^{''(k)} = \downarrow_r^n (Y_{\bullet,i}^{''(k-1)} \otimes w) \quad k = 1, 2, \ldots , \tag{1}$$

$\downarrow_r^n$ denotes down-sampling with reduction factor n and $\otimes$ is the convolution operation. The Laplacian pyramid $Y_r^{'(k)}$ contains band pass components and provides a good approximation to the Laplacian of the Gaussian kernel. It can be constructed by differencing single Gaussian pyramid layers:

$$Y_r^{'(k)} = Y_r^{''(k)} - \uparrow_r^n (Y_{\bullet}^{''(k+1)}), \quad k = 0, 1, \ldots . \tag{2}$$

Each resolution data are independently modelled by their dedicated SAR.

3   SAR Factorial Texture Model

Single multispectral smooth texture factors are modelled using the multispectral simultaneous autoregressive model (SAR) [1]. The 3D SAR model relates each zero mean pixel value Y_r by a linear combination of neighbouring pixel values and an additive uncorrelated Gaussian noise component[1]:

$$Y_{r,i} = \sum_{j=1}^{d} \sum_{s \in I_r^{i,j}} a_{s,i,j} Y_{r \oplus s,j} + e_{r,i}, \qquad i = 1, \ldots, d , \tag{3}$$

where d equals the number of image spectral planes, $I_r^{i,j}$ denotes the neighbour set relating pixels in plane i to neighbouring ones in plane j, $a_{s,i,j}$, $s \in I_r$ are the corresponding parameters which define the dependence of $Y_{r,i}$ on its neighbour sets $I_r^{i,j} \, \forall j$. The driving Gaussian noise $e_{r,i}$ are i.i.d. random variables with zero mean and the i−th spectral plane variance is denoted σ_i. The symbol

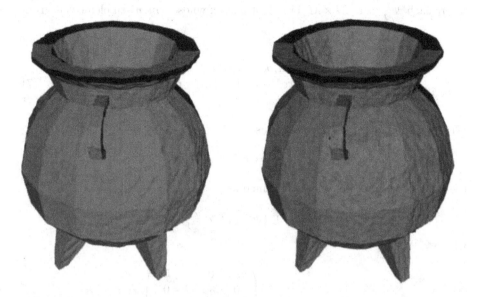

Fig. 2. Synthetic leather (left) and foil mapped to the kettle object

$\oplus$ denotes modulo M addition in each index. Note that the SAR model can be easily defined also for other than the Gaussian noise.

Rewriting the autoregressive equation (3) to the matrix form for the multispectral model, i.e., $i \in \{1, \ldots, d\}$, the SAR model equations are

$$\Psi Y = \epsilon \tag{4}$$

where

$$\Psi = \begin{pmatrix} \Psi_{11} & \Psi_{12} & \ldots & \Psi_{1d} \\ \Psi_{21} & \Psi_{22} & \ldots & \Psi_{2d} \\ \vdots & \vdots & & \vdots \\ \Psi_{d1} & \Psi_{d2} & \ldots & \Psi_{dd} \end{pmatrix}, \tag{5}$$

$$Y = [Y_{[1]}, Y_{[2]}, \ldots, Y_{[d]}]^T ,$$
$$\epsilon = [e_{[1]}, e_{[2]}, \ldots, e_{[d]}]^T ,$$

and both $Y_{[i]}$ and $e_{[i]}$ are M^2 vectors of lexicographic ordered arrays $\{Y_{\bullet,i}\}$ and $\{e_{\bullet,i}\}$. The transformation matrix Ψ is composed of $M^2 \times M^2$ block circulant submatrices (6):

$$\Psi_{ij} = \begin{pmatrix} \Psi_{ij}^1 & \Psi_{ij}^2 & \ldots & \Psi_{ij}^M \\ \Psi_{ij}^M & \Psi_{ij}^1 & \ldots & \Psi_{ij}^{M-1} \\ \vdots & \vdots & \ddots & \vdots \\ \Psi_{ij}^2 & \Psi_{ij}^3 & \ldots & \Psi_{ij}^1 \end{pmatrix} \tag{6}$$

where each Ψ_{ij}^k is a $M \times M$ circulant matrix whose (m, n)-th element is given by:

$$\Psi_{ij,m,n}^k = \begin{cases} 1, & i = j, \ m = n, \ k = 1, \\ -a_{s,i,j}, \ s_1 = k - 1, & s_2 = ((n - m) \mod M), \ (s_1, s_2) \in I_r^{ij}, \\ 0, & otherwise. \end{cases} \quad (7)$$

Writing the image observations (4) as

$$Y = \Psi^{-1} \epsilon \ ,$$

the image covariance matrix is obtained as

$$\Sigma_Y = E\{YY^T\} = E\left\{\Psi^{-1} \epsilon \, \epsilon^T \Psi^{-T}\right\} = \Psi^{-1} \Sigma_\epsilon \, \Psi^{-T}$$

where

$$\Sigma_\epsilon = E\{\epsilon\epsilon^T\} = \begin{pmatrix} \sigma_1 I & 0 & \dots & 0 \\ 0 & \sigma_2 I & \dots & 0 \\ \vdots & & & \vdots \\ 0 & 0 & \dots & \sigma_d I \end{pmatrix}. \quad (8)$$

3.1   Parameter Estimation

The selection of an appropriate SAR model support is important to obtain good results in modelling of a given random field. If the contextual neighbourhood is too small, it cannot capture all details of the random field. Contrariwise, inclusion of the unnecessary neighbours adds to the computational burden and can potentially degrade the performance of the model as an additional source of noise. Direct selection of the optimal support requires numerical optimization hence we exploit a spatial correlation approach [15]. Similarly, both Bayesian as well as the maximum likelihood SAR parameter estimators require numerical optimization.

A least squares (LS) SAR model parameters estimate allows to avoid an expensive numerical optimization method at the cost of accuracy. It can be obtained by equating the observed pixel values of an image to the expected value of the model equations. For a multispectral SAR model this task leads to d independent systems of M^2 equations:

$$Y_{r,i} = E\{Y_{r,i} \mid \gamma_i\} = X_{r,i}^T \gamma_i, \qquad r \in I, \ i \in \{1, \dots, d\} \ , \quad (9)$$

$$\gamma_i = [\gamma_{i1}, \gamma_{i2}, \dots, \gamma_{id}]^T \ ,$$

$$X_{r,i} = \left[\{Y_{r \oplus s,1} : s \in I_r^{i1}\}, \{Y_{r \oplus s,2} : s \in I_r^{i2}\}, \dots, \{Y_{r \oplus s,d} : s \in I_r^{id}\}\right]^T \ ,$$

where $\gamma_{ij} = [a_{s,i,j} : \forall s \in I_r^{i,j}]$ and for which the LS estimates $\hat{\gamma}_i$ and $\hat{\sigma}_i$ can be found as

$$\hat{\gamma}_i = \left(\sum_{s \in I} X_{s,i} X_{s,i}^T \right)^{-1} \left(\sum_{s \in I} X_{s,i} Y_{s,i} \right) ,$$

$$\hat{\sigma}_i = \frac{1}{M^2} \sum_{s \in I} \left(Y_{s,i} - \hat{\gamma}_i^T X_{s,i} \right)^2 .$$

3.2 SAR Model Synthesis

A general multidimensional SAR model has to be synthesized using some of the Markov Chain Monte Carlo (MCMC) methods. Due to our toroidal lattice assumption we can use a noniterative efficient synthesis which uses the discrete fast Fourier transformation (DFT) instead. The SAR model equations (3) may be expressed in terms of the DFT of each image plane as:

$$\tilde{Y}_{t,i} = \sum_{j=1}^{d} \sum_{s \in I_r^{ij}} a_{s,i,j} \, \tilde{Y}_{t,j} \, e^{\sqrt{-1}\omega_{st}} + \tilde{\epsilon}_{t,i}, \qquad i = 1, \ldots, d, \qquad (10)$$

where $\tilde{Y}_{t,i}$ and $\tilde{\epsilon}_{t,i}$ are the 2D DFT coefficients of the image observation and noise sequences $\{Y_{s,i}\}$ and $\{e_{s,i}\}$, respectively, at discrete frequency index $t = (m,n)$ and $\omega_{rt} = \frac{2\pi(mr_1 + nr_2)}{M}$. For the multispectral model this can be written in matrix form as

$$\tilde{Y}_t = \Lambda_t^{-1} \Sigma^{\frac{1}{2}} \tilde{\epsilon}_t, \qquad t \in I, \qquad (11)$$

where

$$\tilde{Y}_t = [\tilde{Y}_{t,1}, \tilde{Y}_{t,2}, \ldots, \tilde{Y}_{t,d}]^T ,$$
$$\tilde{\epsilon}_t = (\tilde{\epsilon}_{t,1}, \tilde{\epsilon}_{t,2}, \ldots, \tilde{\epsilon}_{t,d})^T ,$$

$$\Sigma^{\frac{1}{2}} = \begin{pmatrix} \sqrt{\sigma_1} & 0 & \cdots & 0 \\ 0 & \sqrt{\sigma_2} & \cdots & 0 \\ \vdots & & & \vdots \\ 0 & 0 & \cdots & \sqrt{\sigma_d} \end{pmatrix} ,$$

$$\Lambda_t = \begin{pmatrix} \lambda_{t,11} & \lambda_{t,12} & \cdots & \lambda_{t,1d} \\ \lambda_{t,21} & \lambda_{t,22} & \cdots & \lambda_{t,2d} \\ \vdots & & & \vdots \\ \lambda_{t,d1}(t) & \lambda_{t,d2} & \cdots & \lambda_{t,dd} \end{pmatrix} ,$$

$$\lambda_{t,ij} = \begin{cases} 1 - \sum_{s \in I_r^{ij}} a_{s,i,j} \, e^{\sqrt{-1}\omega_{st}} & i = j , \\ -\sum_{s \in I_r^{ij}} a_{s,i,j} \, e^{\sqrt{-1}\omega_{st}} & i \neq j . \end{cases}$$

The SAR model is stable and valid if Λ_t is non-singular matrix $\forall t \in I$. Given the estimated model parameters, a $d \times M \times M$ multispectral SAR image can be non iteratively synthesized using the following algorithm:

1. Generate the i.i.d. noise arrays $\{e_{r,i}\}$ for each image plane using a Gaussian random number generator.
2. Calculate the 2D DFT of each noise array, i.e., produce the transformed noise arrays $\{\tilde{\epsilon}_{t,i}\}$.
3. For each discrete frequency index t, compute $\tilde{Y}_t = \Lambda_t^{-1} \Sigma^{\frac{1}{2}} \tilde{\epsilon}_t$.
4. Perform the 2D inverse DFT of each frequency plane $\{\tilde{Y}_{t,i}\}$, producing the synthesized image planes $\{Y_{s,i}\}$.

The resulting image planes will have zero mean thus it is necessary to add the estimated mean to each spectral plane in the end. The fine resolution texture is obtained from the pyramid collapse procedure (inversion process to process described in section 2.2).

4   Results

We have tested the BTF-SAR model algorithm on BTF colour textures from the University of Bonn BTF measurements [17,21] (among several available materials are leather, wood, or wool). Each BTF material sample comprised in the University of Bonn database is measured in 81 illumination and 81 viewing angles and has resolution 800×800 pixels, so that 81×81 images had to be analysed for each material. Fig.2 demonstrates the synthesised result for leather and foil materials, i.e. synthesised BTF textures combined with their range maps in the displacement mapping filter of the rendering Blender[1] software with the

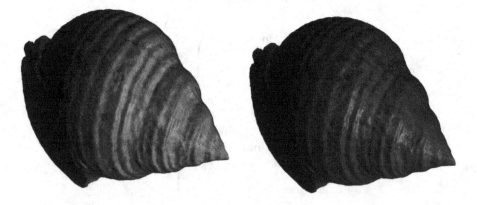

Fig. 3. Two types of modelled wood textures mapped to the conch model

---

[1] www.blender.org

BTF plugin [19,20] and mapped to the kettle model[2] The scene was rendered in several different illumination conditions to demonstrate visual quality of the synthesised BTF. The same approach for two different wood varieties and a detailed conch model measured using our Konika-Minolta laser scanner is illustrated on Fig.3.

Table 1. Mutual comparison of 3D Gaussian Markovian models, PC with 2 Dual Core Pentium 2.8 GHz CPU time was tested using the symmetric half of the first order hierarchical neighbourhood (i.e. 3 neighbours) on 800×800 training and 512×512 synthesized texture, respectively

model	Markovianity	analysis	synthesis	support
			Properties	
SAR	wide-sense	approx.	MCMC / FFT	general
CAR	wide-sense	analytical	analytical	causal / unilateral
GMRF	strict-sense	approx.	MCMC / FFT	symm. hierarchical

model	stability	analysis	synthesis
		CPU time [s]	
SAR	+	0.25	1.35
CAR	−	0.55	0.15
GMRF	+	3.9	2.5

Tab.1 surveys the basic features of three related SAR, CAR [12], GMRF [10] factorial texture models which can alternate in the overall BTF model. All these 3D models can be expressed in the autoregressive form but only the CAR model can be solved analytically [12], does not require the toroidal border assumption, and has by the order of magnited faster synthesis. However, both non causal models (SAR, GMRF) are more general and robust. Analogous conclusions hold also for their 2D counterparts.

5   Conclusion

Our testing results of the algorithm on all available BTF data are promising although they are based on the mathematical model in contrast to the intelligent sampling approach, and as such they can only approximate realism of the original measurement. Some synthetic textures reproduce given measured texture images so that both natural and synthetic textures are almost visually indiscernible. Even the not so successful results can be used for the preattentive BTF textures applications. The main benefit of this inherently multispectral method is more realistic representation of texture colourfulness, which is naturally apparent in

---

[2] http://e2-productions.com/repository/modules/PDdownloads/singlefile.php?cid=10&lid=388

case of very distinctively coloured textures. The multi scale approach is more robust and sometimes allows better results than the singlescale one.

The presented BTF-SAR model offers fast seamless enlargement of BTF texture to arbitrary size, very high BTF texture compression ratio which cannot be achieved by any other sampling based BTF texture synthesis method. This is advantageous for transmission, storing or modelling visual surface texture data while the model has still moderate computation complexity. The method does not need any time consuming numerical optimisation like the usually employed Markov chain Monte Carlo method or some of their deterministic approximation. In addition, this model may be used to reconstruct BTF space (i.e. missing parts of the BTF measurement space) or even non existing (i.e. previously not measured or edited) BTF textures. The model is also potentially capable of direct implementation inside the graphical card processing unit or a multithreaded implementation.

Acknowledgements. This research was supported by the grant GAČR 102/08/0593 and partially by the projects MŠMT 1M0572 DAR, GAČR 103/11/0335, CESNET 387/2010.

References

1. Bennett, J., Khotanzad, A.: Multispectral random field models for synthesis and analysis of color images. IEEE Trans. on Pattern Analysis and Machine Intelligence 20(3), 327–332 (1998)
2. Blinn, J.: Simulation of wrinkled surfaces. SIGGRAPH 1978 12(3), 286–292 (1978)
3. Dana, K.J., Nayar, S.K., van Ginneken, B., Koenderink, J.J.: Reflectance and texture of real-world surfaces. In: CVPR, pp. 151–157. IEEE Computer Society (1997)
4. De Bonet, J.: Multiresolution sampling procedure for analysis and synthesis of textured images. In: ACM SIGGRAPH 1997, pp. 361–368. ACM Press (1997)
5. Efros, A.A., Freeman, W.T.: Image quilting for texture synthesis and transfer. In: Fiume, E. (ed.) ACM SIGGRAPH 2001, pp. 341–346. ACM Press (2001), citeseer.nj.nec.com/efros01image.html
6. Favaro, P., Soatto, S.: 3-D shape estimation and image restoration: exploiting defocus and motion blur. Springer-Verlag New York Inc. (2007)
7. Filip, J., Haindl, M.: Bidirectional texture function modeling: A state of the art survey. IEEE Transactions on Pattern Analysis and Machine Intelligence 31(11), 1921–1940 (2009)
8. Frankot, R.T., Chellappa, R.: A method for enforcing integrability in shape from shading algorithms. IEEE Trans. on Pattern Analysis and Machine Intelligence 10(7), 439–451 (1988)
9. Haindl, M.: Texture synthesis. CWI Quarterly 4(4), 305–331 (1991)
10. Haindl, M., Filip, J.: Fast BTF texture modelling. In: Chantler, M. (ed.) Proceedings of Texture 2003, pp. 47–52. IEEE Press, Edinburgh (2003)
11. Haindl, M., Filip, J.: A Fast Probabilistic Bidirectional Texture Function Model. In: Campilho, A.C., Kamel, M.S. (eds.) ICIAR 2004, Part II. LNCS, vol. 3212, pp. 298–305. Springer, Heidelberg (2004)

12. Haindl, M., Filip, J., Arnold, M.: BTF image space utmost compression and modelling method. In: Kittler, J., Petrou, M., Nixon, M. (eds.) Proceedings of the 17th IAPR International Conference on Pattern Recognition, vol. III, pp. 194–197. IEEE, Los Alamitos (2004), http://dx.doi.org/10.1109/ICPR.2004.1334501

13. Haindl, M., Hatka, M.: BTF Roller. In: Chantler, M., Drbohlav, O. (eds.) Proceedings of the 4th International Workshop on Texture Analysis, Texture 2005, pp. 89–94. IEEE, Los Alamitos (2005)

14. Haindl, M., Hatka, M.: A roller - fast sampling-based texture synthesis algorithm. In: Skala, V. (ed.) Proceedings of the 13th International Conference in Central Europe on Computer Graphics, Visualization and Computer Vision, pp. 93–96. UNION Agency - Science Press, Plzen (2005)

15. Haindl, M., Havlíček, V.: Multiresolution colour texture synthesis. In: Dobrovodský, K. (ed.) Proceedings of the 7th International Workshop on Robotics in Alpe-Adria-Danube Region, pp. 297–302. ASCO Art, Bratislava (1998)

16. Haindl, M., Havlíček, V.: A multiscale colour texture model. In: Kasturi, R., Laurendeau, D., Suen, C. (eds.) Proceedings of the 16th International Conference on Pattern Recognition, pp. 255–258. IEEE Computer Society, Los Alamitos (2002), http://dx.doi.org/10.1109/ICPR.2002.1044676

17. Haindl, M., Filip, J.: Advanced textural representation of materials appearance. In: SIGGRAPH Asia 2011 Courses, SA 2011, pp. 1:1–1:84. ACM, New York (2011), http://doi.acm.org/10.1145/2077434.2077435

18. Haindl, M., Havlíček, V.: A Multiresolution Causal Colour Texture Model. In: Amin, A., Pudil, P., Ferri, F., Iñesta, J.M. (eds.) SPR 2000 and SSPR 2000. LNCS, vol. 1876, pp. 114–122. Springer, Heidelberg (2000)

19. Hatka, M.: Btf textures visualization in blender. In: Proceedings of the Graduate Students Days, pp. 37–46. FNSPE CTU (2009)

20. Hatka, M., Haindl, M.: Btf rendering in blender. In: Zhang, X., Pan, Z., Dong, W., Liu, Z.Q. (eds.) Proceedings of the 10th International Conference on Virtual Reality Continuum and Its Applications in Industry, VRCAI 2011, pp. 265–272. ACM, New York (2011), http://doi.acm.org/10.1145/2087756.2087794

21. Müller, G., Meseth, J., Sattler, M., Sarlette, R., Klein, R.: Acquisition, synthesis and rendering of bidirectional texture functions. In: Eurographics 2004. STAR - State of The Art Report, pp. 69–94. Eurographics Association (2004)

22. Wang, L., Wang, X., Tong, X., Lin, S., Hu, S., Guo, B., Shum, H.: View-dependent displacement mapping. ACM Transactions on Graphics 22(3), 334–339 (2003)

23. Woodham, R.: Photometric method for determining surface orientation from multiple images. Optical Engineering 19(1), 139–144 (1980)

Analysis of Human Gaze Interactions
with Texture and Shape

Jiří Filip, Pavel Vácha, and Michal Haindl

Institute of Information Theory and Automation of the ASCR, Czech Republic

Abstract. Understanding of human perception of textured materials is one of the most difficult tasks of computer vision. In this paper we designed a strictly controlled psychophysical experiment with stimuli featuring different combinations of shape, illumination directions and surface texture. Appearance of five tested materials was represented by measured view and illumination dependent Bidirectional Texture Functions. Twelve subjects participated in visual search task - to find which of four identical three dimensional objects had its texture modified. We investigated the effect of shape and texture on subjects' attention. We are not looking at low level salience, as the task is to make a high level quality judgment. Our results revealed several interesting aspects of human perception of different textured materials and, surface shapes.

Keywords: texture, shape, human gaze, perception, psychophysics.

1 Introduction

Many research areas such as computer vision, computer graphics, image understanding, cognitive psychology, start to focus research effort also on human perception of real materials. This research has enormous positive impact on many practical application such as automatic object recognition, scene segmentation, data compression, efficiency of rendering algorithms etc. In this context the best practically available representation of real materials are their view and illumination direction dependent measurements. Such measurements are commonly thought as bidirectional texture functions (BTF) [1]. BTFs represent challenging data due to theirs huge size and thus high processing and rendering expenses [2]. In this paper we used this data to prepare controlled visual search experiment with stimuli preserving realistic appearance of textured materials. In this experiment we investigate the effect of object, surface and texture on overt attention in a demanding visual search task. We are not looking at low level salience, as the task is to make a high level quality judgment. We expect observers to make fixations to regions which they expect to get the most task relevant information from, rather than those which are the most salient.

2 Related Work

In the past visual psychophysics put great effort in understanding a way people perceive and assess objects of different reflectance properties under specific

E. Salerno, A.E. Çetin, and O. Salvetti (Eds.): MUSCLE 2011, LNCS 7252, pp. 160–171, 2012.

illumination conditions. Ho et al. [3] found that roughness perception is corre-
lated with texture contrast. Lawson et al. [4] showed that human performance in
matching 3D shapes is lower for varying view directions. Ostrovsky [5] pointed
out that illumination inconsistency is hard to detect in geometrically irregular
scenes. The importance of specular reflections of environment illuminations for
recognition of 3D shape was analyzed in [7]. A psychophysical analysis of bidirectional reflectance distribution functions (BRDF) dependently on different shapes
and illumination environments was performed in [8], [9],[10]. Basic perceptual
interactions of texture and shape were investigated by Todd and Thaler [11] .

Also human gaze analysis has been widely used in many applications [12]
such as visual search [13], web-sites usability [14], eye motion synthesis [15],
predicting of fixation behavior in computer games [16], assessing visual realism
of medical scenes [17] or quality of BTF resampling [18]. A gaze data analysis
was employed in [19]. Authors analysed correlation of fixations density with of
several statistical texture features. The best features were then used for perceptually driven BTF compression. Although, eye-tracking methods have been used
to analyse the way people perceive or recognize simple 3D objects [20], to the
best of our knowledge no one has analysed human gaze interaction with surface
shape realistically textured by means of bidirectional texture function.

In this paper we build on our recent work [21], where the same experimental
data were used but only subjects responses were analysed, i.e., not gaze fixations. We were comparing abilities of different statistical descriptors of texture
to model subjects' responses. Here we are moving further on and thoroughly
analyse human gaze attention to experimental stimuli.

In the context of multimedia understanding the effective processing of visual
texture information is a must. One way to achieve such an effective processing is analysis of human visual perception and focusing processing algorithm
preferably to visually salient part of the data.

Contribution of the Paper. The main aim of this paper is to explore the effect
of object shape and texture on human gaze attention in a visually complex task
using real textured materials measurements. As we are using one of the most
accurate representation material texture appearance we believe that by taking
into account texture statistics and their interaction with local geometry and
illumination direction, we acquire information about our visual perception of
real materials.

Organization of the Paper. The paper starts with description of psychophysical eye-tracking experiment in Section 3. Subjects' responses withing experiment
are analysed in Section 4, while their gaze data analysis is subject of Section 5.
Main conclusions and future work are outlined in Section 6.

3   Visual Search Experiment

We performed a visual search experiment in order to investigate effects of surface
texture, shape and illumination direction and their interactions.

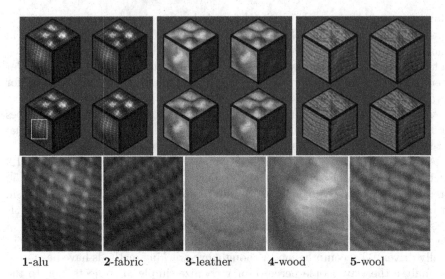

| 1-alu | 2-fabric | 3-leather | 4-wood | 5-wool |

Fig. 1. Examples of the stimuli (first row) and five tested material samples (second row)

Experimental Stimuli. For experimental stimuli we have used static images of size 1000×1000 pixels, featuring four cubes in individual quadrants (first row of Fig. 1). We have used this stimuli layout to avoid the fixations central bias reported in [22], i.e. observers have a tendency to fixate the central area of the screen. The cube faces were modified in a way to feature different geometry on all three visible faces (top, left, right). We used different shape for each cube face: **I**-wide indent (1), **R**-random bumps (2), **B**-wide bump (3), **F**-flat face (4), **H**-horizontal waves (5), **V**-vertical waves (6) as it is show in Fig. 2. For illumination we used directional light from left and right directions parallel with the upper edge of the cubes. This configuration guarantee the same illumination of all cubes in stimuli and similar distribution of light across top and left/right faces in single cubes. Not all combinations of test cube orientations were used in the experiment as this would result in enormous number of stimuli. We used only eleven different orientations selected in a way to allow us to compare the most interesting combinations of faces geometry. Additionally, not all the orientations were illuminated from both directions as shown in Fig. 11. Figure shows orientation number (first row) and shapes of left, right, top faces (third row).

| I | R | B | F | H | V |

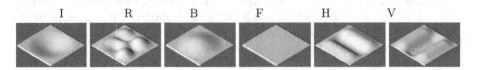

Fig. 2. Tested shapes modifying cube face

Finally, all cubes were rendered using textured materials. We used five different samples of view and illumination dependent textures represented by

Bidirectional Texture Functions (BTF) [23]. In each quadruple three cubes were showing the original data rendering and the remaining one was showing a slightly modified data. We have used three different filters for the original data modification:

A - illumination/view directions downsampling to 50%
B - spatial filtering (averaging by kernel 3×3)
C - spatial filtering (averaging by kernel 5×5).

The proposed filters introduce only very subtle differences (see Fig. 3) between the original and the modified data and force subjects to perform extensive visual search, which allows us to collect detailed gaze data. The edges of the cubes were set to black to mask potentially salient texture seams. Object edges and texture seams are interesting and important sources of visual information, but are deemed out-with the scope of this paper. The background and the remaining space on the screen was set to dark gray. Examples of stimuli and all tested textured materials are shown in Fig. 1.

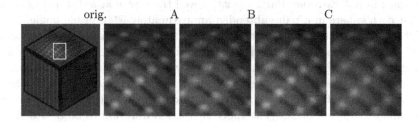

Fig. 3. Performance of the applied filters on sample 1-alu

First row of Fig. 11 shows the 13 conditions of cube orientation and illumination direction that were used. Together with five BTF texture samples, and three different filters, the total number of stimuli was 195 (13×5×3).

Participants. Twelve paid observers (three females, nine males) participated in the experiments. All were students or university employees working in different fields, were less than 35 years of age, and had normal or corrected to normal vision. All were naive with respect to the purpose and design of the experiment.

Experimental Procedure. The participants were shown the 195 stimuli in a random order and asked to identify which of the cubes has a modified/degraded surface texture, i.e., slightly different from the remaining three cubes. Stimulus was shown until one of four response keys, identifying the different cube, was pressed. There was a pause of one second between stimuli presentations, and participants took on average around 90 minutes to perform the whole experiment that was split in four sessions. All stimuli were presented on a

Fig. 4. Setup of the experiment with the eye-tracker highlighted

calibrated 20.1" NEC2090UXi LCD display (60Hz, resolution 1600×1200, color temperature 6500K, gamma 2.2, luminance 120 cd/m^2). The experiment was performed in a dark-room. Participants viewed the screen at a distance of 0.7m, so that each sphere in a pair subtended approximately 9° of visual angle.

Subjects gaze data were recorded in a dark room using a Tobii x50 infrared-based binocular eye-tracking device as shown in Fig. 4-left. The device was calibrated for each subject individually and provided the locations and durations of fixations at a speed 50 samples/s. Maximum error specified by manufacturer is approximately ±0.5^o of visual angle, which corresponds to ±30 pixels for our setup and stimuli resolution. The shortest fixation duration to be recorded was set to 100 ms.

4   Subjects Responses Analysis

First we analysed the subjects ability in finding the modified cube. In average the subjects were able to find the right cube in 67% stimuli, which was surprisingly high given the subtle changes introduced by applied filters (see Fig. 3) (chance level 25%). During an informal interview after experiment subjects mentioned that they were often certain in less than 50% of stimuli and for the rest they were only guessing the right answer. The obtained rates suggests that for difficult cases they often successfully relied on a low level visual perception. The responses accuracy of individual filters is shown in Fig. 5 and reveals that modifications introduced by angular resolution filter **A** are the hardest to spot while the spatial smoothing by filter **B** are the most apparent, as expected since smoothing effect is uniform and generally more apparent in comparison with slight illumination and view direction dependent change in reflectance caused by directions reduction (**A**). While success rates across textures were quite similar for smoothing filters **B** and **C**, their values for filter **A** varied much more. Average values for individual materials are shown in Fig.7-a.

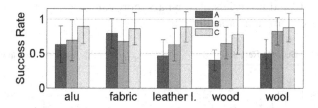

Fig. 5. Percentage of successful responses dependent on data filtration and texture sample

5   Gaze Data Analysis

This section has two parts. In the first, we analyse gaze fixation statistics such as their number, duration etc. across different tested materials. In the second, a spatial distribution of gaze fixation with regards to material and shape is analysed. In both parts a statistics predicting gaze data are offered. In total, twelve subjects performed 62916 fixations longer than 100 ms. Average fixation duration was 242 ms. Each stimulus image was in average observed 11 s and in average comprised 26 fixations.

5.1   Fixation Numbers and Duration Analysis

Fixation duration as a function of ordinal fixation number in Fig. 6-left shows that the average fixation duration was the lowest for the first three fixations and then increased almost linearly with trial duration. This behaviour is similar to results in [13] and suggests that subjects applied a coarse-to-fine approach during visual search. When the difference between cubes is more apparent the subjects notice it within the first few fixations, otherwise they spend more time by detail focused searching for a difference resulting in longer durations of fixations that increase proportionally with total length of the search. Fig. 6-right shows that the total number of fixations decreased almost exponentially and thus most responses to stimuli are given during the first forty fixations.

Figure Fig. 7 presents average success rate obtained from subject keyboard responses (a), image duration (b), number of fixations (c), and fixation duration

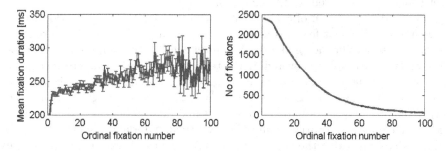

Fig. 6. Mean fixation duration (left) and number of fixations (right) as a function of ordinal fixation number, i.e., average length of first one hundred fixations

(d) for the five tested materials. Error-bars represent twice standard error across different cube orientations and illuminations.

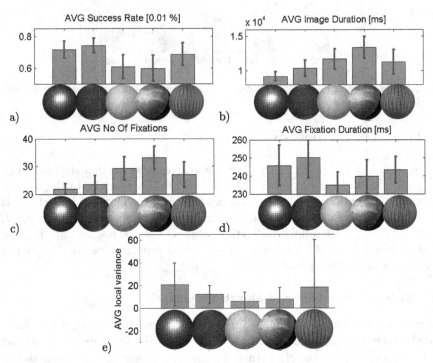

Fig. 7. Average subjects responses success rate (a), average stimulus duration [ms] (b), average number of fixations (c), and average fixation duration [ms] (d), average local variance (kernel 31×31) (e) for the tested material samples

There are apparent differences between materials. For non-regular / more smooth materials **3**-leather and **4**-wood the subjects were less successful in iden-tification of the modified cube, they observed stimuli longer, made significantly more fixations which were much shorter in comparison with the other materials. We assume that when subjects detect regularity in the material, they focus on angular differences only while for non-regular samples they still perform search over wider spatial content of the material.

Motivated by research of Su et al. [24] who reduced texture variation to re-move unwanted salient features and by results from [19], where a texture local variance was identified as predictor of fixation density, we computed average lo-cal variance of the sphere covered by the textures having the same resolution as the stimuli. We used square Gaussian weighted kernel of size 31×31 pixels. This setting roughly corresponds in our setup to ±0.5° of visual angle and thus the kernel should comprise approximately the same information as a focused eye. Values of average local variance filter for the tested material are shown in Fig. 7-e. Error-bars are twice the standard deviation of local variances across whole image. From the results it is apparent that the lower is average local variance of

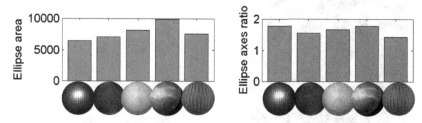

Fig. 8. Areas of fixation cloud fitted ellipses (left) and their axes ratio (right) for individual texture samples

the material, (1) the more time and fixations subjects need to carry on with a given task and (2) the subjects' success rate is lower as there more likely prevail only low frequency features, which is the case of smooth materials (**3**, **4**). For these materials it is more difficult for subjects to recognize the effect of applied filter. This conclusion was supported also by fitting PCA to fixations clouds for individual materials and representing the two principal components as axes of fixation ellipse. The area of such an ellipse for different materials is shown in Fig. 8-left. This suggests that non-regular/smoother textures, receive more but short fixations and the search area is wider. Fig. 8-right shows dependency of ratio of ellipse axes length on material sample (i.e., for 1 = circle). This ratio is higher for more glossy materials (**1**-alu, **3**-leather, **4**-wood) than for the other more diffuse materials. It might suggests that fixation cloud is shaped according specular reflection or other locally salient features of the sample and thus perhaps better follows places with important visual information.

5.2   Spatial Analysis of Fixations

Next a spatial analysis of gaze fixations was carried out. A typical search pattern is shown in Fig. 9-left. We can see that subject tended to do horizontal and vertical saccades (eye moves between fixations) in square pattern, while spending just one or two fixations on each cube. This fact is proved by histogram of a number of fixations on the same cube before jump to any other Fig. 9-right. As follows also from further analysis subjects very rarely made a diagonal saccade, which might be caused by the fact that in real life situations the majority of important visual information is organized either horizontally or vertically.

Interesting results revealed fixation analysis across different cube faces (shapes). Fig. 10-a shows such a distributions of saccades and fixations across different shapes Fig. 2, while Fig. 10-b shows avrage fixation duration. Surface with curvature (1-indent (I), 2-random bumps (R), 3-bump (B), 5,6-waves (V,H)) received more fixations while subjects were not interested in shape 4-flat surface. This fact can be reasoned by presence of less information on such a flat face in comparison with other shapes, which subjects tend to use more frequently.

We have found that distribution of attention across different shapes can be again relatively reliably predicted by the average local variance (Gaussian weighted kernel 31×31) as shown in Fig. 10-c. Interestingly, the number of

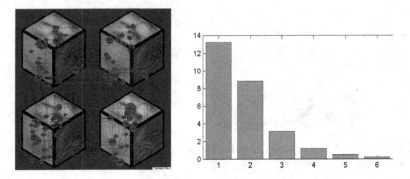

Fig. 9. Example of a typical search pattern (left), histogram of a number of fixations on the same cube before jump to any other (right)

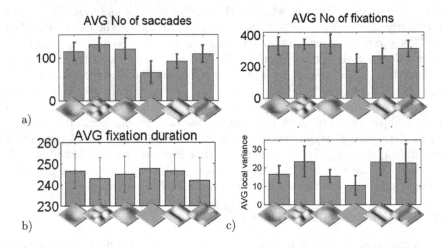

Fig. 10. Comparison of normalized numbers of saccades (left) and fixations (right) (a) and their average duration (b). Prediction based on average local variance (kernel 31×31) for individual cube faces (c).

fixations and saccades (Fig. 10-a) is inversely proportional their average duration (Fig. 10-b), i.e., the more fixations the shape receives the more information it conveys, and thus less time is required per each fixation.

Subjects attention to individual shapes illuminated from two different directions is shown in Fig. 11. The more thick is a dot/line in the second row of images the more fixations/saccades corresponding face or transitions between faces receives in average. Surprisingly, it is not always the case that the most fixations is focused on the two sides of a cube which are illuminated (Fig. 1). In the case when a flat face is illuminated it receives less fixations than shadowed face (i.e., poses 1-L, 2-R, 3-L). This suggests that subjects prefer to observe curved texture rather than uniform flat texture during their task even though such texture is dimly illuminated, i.e., has lower contrast. This conforms with findings of Mackworth and Morandi [25] showing that gaze selects informative details in the pictures.

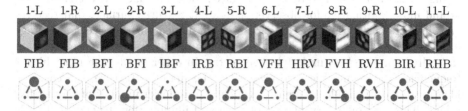

Fig. 11. Geometry significantly influences human gaze attention to texture. Intra-cube (second row) saccades and fixations for all tested cube poses (first row). Thickness of lines (inter-face saccades) and dots (intra-face fixations) represent a number of saccades/fixations.

Similar data across all cube orientations are visualized in the second row of Fig. 12, but there are additional plots in the first row, showing similarly average fixations/saccades between individual cubes in stimuli. It is apparent

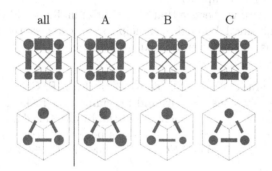

Fig. 12. Inter-cube (first row) and intra-cube (second row) saccades across all stimuli (left) and for individual filters. Thickness of lines (inter-cube/face saccades) and dots (intra-cube/face saccades) correlates with a number of saccades.

that subjects tend to avoid diagonal saccades between cubes and the upper cubes receive more of the saccades/fixations. Fig. 12 also compares gaze attention response to different data modification filters (**A,B,C**) and supports our previous conclusion that modification by filter **A** (angular downsampling) is the most difficult to spot as the number of fixations/saccades is similar regardless the position in the stimuli, and thus subjects are forced to perform extensive visual search.

6   Conclusions and Future Work

This paper provides an psychophysical analysis of human gaze attention to textured materials and its interaction with surface shape. It revealed several interesting facts. First, a shaped textured surface is definitely more attractive to look at and more informative for observer than a flat textured surface, that receives only half of the fixations in comparison with shaped surface. Second, average local variance of a curved surface texture can predict observers' gaze attention to both texture and its underlying geometry. In other words, the more higher frequencies and regularities are present in the material texture, the easier is identification of possible differences, which require lower number of shorter fixations. Third, angular degradation of view and illumination dependent data is material dependent and less apparent in comparison with a very subtle data spatial smoothing . Finally, upper parts of stimuli receive generally more attention mainly at the beginning of the trial.

Among applied contexts of this work are fast texture data search and retrieval algorithms predicting human attention by means of local variance. These algorithms can exploit the fact that observers tend to seek information in horizontal and vertical directions with more attention to information presented on non-planar shapes in the first half of shown visual frame. Another application is effective perceptually driven compression and rendering of texture data in virtual and mixed-reality applications.

As our future work we believe that these initial findings and their further investigation in multi-modal domain will help to improve performance of methods processing and visualizing accurate texture data.

Acknowledgements. We would like to thank Prof. P. Green, Dr. A. Clarke, Prof. M. Chantler for consultations and providing eye-tracking device, all subjects participating in the experiments, and the Bonn BTF database for providing their BTF measurements. This work has been supported by GAČR grants 103/11/0335 and 102/08/0593, EC Marie Curie ERG 239294, and CESNET grants 378 and 409.

References

1. Dana, K., van Ginneken, B., Nayar, S., Koenderink, J.: Reflectance and texture of real-world surfaces. ACM Transactions on Graphics 18(1), 1–34 (1999)
2. Filip, J., Haindl, M.: Bidirectional texture function modeling: A state of the art survey. IEEE Transactions on Pattern Analysis and Machine Intelligence 31(11), 1921–1940 (2009)
3. Ho, Y., Landy, M., Maloney, L.: Conjoint measurement of gloss and surface texture. Psychological Science 19(2), 196–204 (2008)
4. Lawson, R., Bülthoff, H., Dumbell, S.: Interactions between view changes and shape changes in picture - picture matching. Perception 34(12), 1465–1498 (2003)
5. Ostrovsky, Y., Cavanagh, P., Sinha, P.: Perceiving illumination inconsistences in scenes. Perception 34, 1301–1314 (2005)

6. Fleming, R.W., Dror, R.O., Adelson, E.H.: Real-world illumination and the perception of surface reflectance properties. Jorunal of Vision (3), 347–368 (2003)
7. Fleming, R.W., Torralba, A., Adelson, E.H.: Specular reflections and the perception of shape. Journal of Vision 4(9) (2004)
8. Vangorp, P., Laurijssen, J., Dutre, P.: The influence of shape on the perception of material reflectance 26(3) (2007)
9. Ramanarayanan, G., Ferwerda, J., Walter, B., Bala, K.: Visual equivalence: Towards a new standard for image fidelity, 26(3) (2007)
10. Křivánek, J., Ferwerda, J.A., Bala, K.: Effects of global illumination approximations on material appearance. ACM Trans. Graph. 29, 112:1–112:10 (2010)
11. Todd, J., Thaler, L.: The perception of 3d shape from texture based on directional width gradients. Journal of Vision 10(5), 17 (2010)
12. Duchowski, A.T.: A breadth-first survey of eye-tracking applications. Behav. Res. Methods Instrum. Comput. 34(4), 455–470 (2002)
13. Over, E., Hooge, I., Vlaskamp, B., Erkelens, C.: Coarse-to-fine eye movement strategy in visual search. Vision Research 47(17), 2272–2280 (2007)
14. Nielsen, J., Pernice, K.: Eyetracking Web Usability. Voices That Matter. New Riders (2009)
15. Deng, Z., Lewis, J.P., Neumann, U.: Automated eye motion using texture synthesis. IEEE Computer Graphics and Applications 25(2), 24–30 (2005)
16. Sundstedt, V., Stavrakis, E., Wimmer, M., Reinhard, E.: A psychophysical study of fixation behavior in a computer game. In: APGV 2008, pp. 43–50. ACM (2008)
17. Elhelw, M.A., Nicolaou, M., Chung, J.A., Yang, G.Z., Atkins, M.S.: A gaze-based study for investigating the perception of visual realism in simulated scenes. ACM Transactions on Applied Perception 5(1) (2008)
18. Filip, J., Chantler, M., Haindl, M.: On uniform resampling and gaze analysis of bidirectional texture functions. ACM Transactions on Applied Perception 6(3), 15 (2009)
19. Filip, J., Haindl, M., Chantler, M.: Gaze-motivated compression of illumination and view dependent textures. In: Proceedings of the 20th International Conference on Pattern Recognition (ICPR), pp. 862–864 (August 2010)
20. Leek, E.C., Reppa, I., Rodriguez, E., Arguin, M.: Surface but not volumetric part structure mediates three-dimensional shape representation: Evidence from part-whole priming. The Quarterly Journal of Experimental Psychology 62(4), 814–830 (2009)
21. Filip, J., Vácha, P., Haindl, M., Green, P.R.: A Psychophysical Evaluation of Texture Degradation Descriptors. In: Hancock, E.R., Wilson, R.C., Windeatt, T., Ulusoy, I., Escolano, F. (eds.) SSPR & SPR 2010. LNCS, vol. 6218, pp. 423–433. Springer, Heidelberg (2010)
22. Tatler, B.W.: The central fixation bias in scene viewing: Selecting an optimal viewing position independently of motor biases and image feature distributions. Journal of Vision 7(14), 1–17 (2007)
23. Database BTF, Bonn (2003), http://btf.cs.uni-bonn.de
24. Su, S., Durand, F., Agrawala, M.: De-emphasis of distracting image regions using texture power maps. In: APGV 2005: 2nd Symposium on Applied Perception in Graphics and Visualization (2005)
25. Mackworth, N., Morandi, A.: The gaze selects informative details within pictures. Attention, Perception, & Psychophysics 2, 547–552 (1967) 10.3758/BF03210264

Rich Internet Application for Semi-automatic Annotation of Semantic Shots on Keyframes

Elisabet Carcel, Manuel Martos, Xavier Giró-i-Nieto,
and Ferran Marqués

Technical University of Catalonia (UPC),
Jordi Girona 1-3, Barcelona, Catalonia, Spain
{xavier.giro,ferran.marques}@upc.edu
http://www.upc.cat

Abstract. This paper describes a system developed for the semi-automatic annotation of keyframes in a broadcasting company. The tool aims at assisting archivists who traditionally label every keyframe manually by suggesting them an automatic annotation that they can intuitively edit and validate. The system is valid for any domain as it uses generic MPEG-7 visual descriptors and binary SVM classifiers. The classification engine has been tested on the multiclass problem of semantic shot detection, a type of metadata used in the company to index new content ingested in the system. The detection performance has been tested in two different domains: *soccer* and *parliament*. The core engine is accessed by a Rich Internet Application via a web service. The graphical user interface allows the edition of the suggested labels with an intuitive drag and drop mechanism between rows of thumbnails, each row representing a different semantic shot class. The system has been described as complete and easy to use by the professional archivists at the company.

Keywords: annotation, RIA, semantic shot, classification, MPEG-7 visual descriptors.

1 Motivation

This paper describes the creation of a web interface for keyframe-based annotation of semantic shots. A semantic shot provides information about the camera shot size in terms of field of view (long shot, medium shot, close-up...) plus a subject in the scene, as for example the presence of a player, the sports field or the audience in a soccer match. Figure 1 shows a visual representation of the semantic shot concept.

This semantic shot information is used by archivists at the Catalan Broadcast Corporation (CCMA) to generate a storyboard that describes a video sequence for its indexing and posterior retrieval. The presented tool reduces the necessary time to generate semantic shot metadata by assisting these archivists in their work.

Previous work has studied the automatic detection of semantic shot types. Wang et al [1] worked with three different shot sizes for any generic domain.

E. Salerno, A.E. Çetin, and O. Salvetti (Eds.): MUSCLE 2011, LNCS 7252, pp. 172–182, 2012.

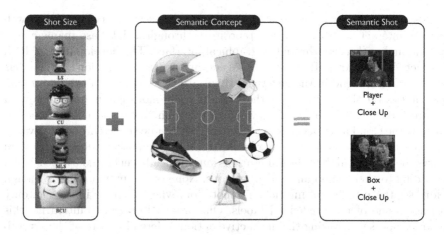

Fig. 1. Semantic shot concept

In their proposal, video shots are uniformly sampled to extract those frames to be analysed. Two types of feature vectors are used, one corresponding to global features such as color, edge and texture and the other one to local features involving the three largest regions and their features. Each vector of features is sent to a different classifier (global and local) to decide which shot size suits best the image among the three predefined classes. The classifier results are fused with a function that combines product and square root. Duan et al [2] presented a framework for semantic shot classification of sports video. The proposed scheme predefined a catalogue of semantic shot classes for tennis, soccer and basketball that covers most of the sports broadcasting video. The system makes use of the domain knowledge (in this case, a specific sport between tennis, soccer or basketball) to perform the video shot classification. For a given domain they construct middle level features (camera motion patterns, motion entropy, shot pace, active regions, etc.) from low level features. The relation between low-level and middle-level is defined by supervised learning methods. Once the middle-level shot attributes are defined, video shots are classified into the predefined classes using Bayesian classifiers and support vector machines. Our work provides similar functionalities with a much simpler scheme, as it does not require local nor mid-level features. The presented solution is only based on standardized and generic MPEG-7 descriptors, each of them to train a separate classifier whose predictions will be fused with an additional late fusion classifier. Moreover, our scheme also considers the possibility of a keyframe being assigned to an additional *clutter* class.

The digitalization of video archives during the last decades together with the evolution of networking technologies boosted by the Internet growth, have driven to the convergence of both worlds. Nowadays video content represents the highest volume of data content transported through the Internet, and web-based technologies are being adopted for accessing video repositories. This philosophy also makes that additional plug-ins to the video archive, as the one presented in this paper, are to be integrated to the system as web services. These web services interact with the content repository through a database management system, and with users through a graphical interface. The development of Rich Internet Applications (RIAs) for accessing automatic annotation systems has been a growing trend in the last years thanks to the maturity of the interaction possibilities on the browser together with asynchronous communications with remote servers. Borth et al [4] developed a system that uses online videos to train on the cloud a set of concept classifiers that can be tuned through relevance feedback from the user. In the work by Bertini et al [3], another web-based system allows the user to validate the annotations automatically generated by the system according to their audio and video features. None of the previous cases though addresses the problem of multiclass annotation, where a single image can only belong to one of the predefined labels. Our proposal offers a solution for this annotation case exploiting the interactive options offered by state of the art web technologies.

The paper is structured in two main parts. Section 2 focuses on the keyframe classifier and the performance results obtained in two different broadcast domains. Section 3 presents the developed Rich Internet Application (RIA) that allows exploiting the classification engine from a web-based graphical user interface. Finally, Section 4 summarizes the main contributions of this work and points at future research directions.

2   Semantic Shot Detection

The pattern recognition part of the system runs on a server that accesses the content repository through a local network. The engine predicts the semantic shot type of those keyframes that are to be annotated. These keyframes have been automatically extracted when the new video shots are ingested in the repository. The keyframe extractor was already integrated in the broadcaster architecture and it was out of the scope of this work to modify it. The generic detection engine has been tested in two different domains: *soccer* and the *parliament*.

2.1   Requirements

The archivists at the broadcast company have their own ontology where the semantic shot types of interest are defined. These semantic classes are particular for this archive but share the main guidelines used by similar companies. The developed system has always worked with real annotations provided by these archivists, but the whole solution is applicable to any other similar domain.

In the *soccer* domain, most of the keyframes showing the playground lack any interest for archivists. These general shots must be distinguished from those of interest, such as the audience or the stadium overview. Other relevant semantic shots are close-ups showing a specific player or people in the VIP box. The six different types of semantic shots defined for the soccer domain are depicted in Fig. 2. An additional *clutter* class represents the non-interesting shot types.

Soccer match Ontology	
Class 1	Player Medium Shot
Class 2	Player Close-Up
Class 3	Stadium Overview
Class 4	Audience
Class 5	Banner
Class 6	Box
Class 7	Clutter

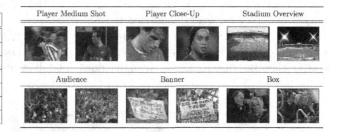

Fig. 2. Soccer shot types

On the other hand, the *parliament* domain is a controlled environment where camera location and shot sizes are normally fixed. This makes it easy to extract event information out of the shot type used in each keyframe. Fig. 3 contains the list and examples of the semantic shot types defined for this domain.

Catalan Parliament Ontology	
Class 1	President Close-Up
Class 2	Close-Up
Class 3	Medium Shot
Class 4	General Chamber Shot
Class 5	General Bureau Shot
Class 6	Clutter

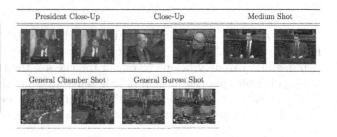

Fig. 3. Parliament shot types

2.2 Design

The proposed system is based around a detector that estimates the probability of an image of belonging to a certain class. This prediction is based on a set of four automatically extracted image features defined in the MPEG-7 standard [6]: Dominant Color, Color Structure, Color Layout and Edge Histogram. The detection of a semantic shot type follows the traditional scheme for supervised learning. A labelled training dataset is processed to extract the visual features from each of its keyframes. The labelled features are analysed by a learning algorithm that trains an SVM classifier [7] to determine the decision boundary between the two basic classes (positive and negative). When an new keyframe is to be annotated, its features are compared with the semantic shot model to estimate the probability of the keyframe of belonging to the class. A final decision about the detection of the semantic shot type is taken by comparing the probability with a predefined threshold.

The addressed problem though is not just a binary detection but a multiclass prediction, as every keyframe can only belong to one of the defined classes.

The binary scheme has been adapted into a multiclass scenario by training a binary classifier for every modelled semantic shot type (not the clutter class) and, at prediction time, deciding the class with the highest probability. The whole process is depicted in Figure 4. Notice that no classifier is trained for the *clutter* class because its visual diversity does not follow any predictable pattern.

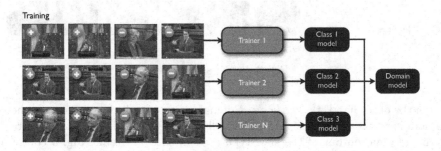

Fig. 4. Training architecture

At the detection stage shown in Fig. 5, each semantic shot detector will produce a score based on the visual descriptors of the unlabelled keyframes. As each instance can only belong to one of the classes, the maximum score among them will determine the most probable semantic shot. The maximum score is processed by a decision block, where a semantic shot label if over the detection threshold. Otherwise, the keyframe will be classified as *clutter*.

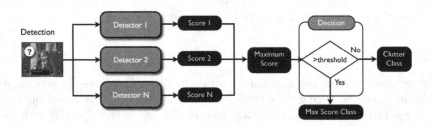

Fig. 5. Detection architecture

2.3   Evaluation

The system has been tested by tuning three different parameters that configure the trainer and the detector. Since every class may be formed by several distinctive views, a different classifier is trained for each view. Each of these view is determined by means of an unsupervised clustering algorithm. In our work, the *Quality Threshold* solution [5] is adopted, an algorithm that requires two configuration parameters: the minimum amount of elements necessary to define a cluster, and the maximum radius of a cluster. Thanks to this step, the view diversity within a semantic shot class can be estimated and learned separately. Additional, the detector also requires the definition of the minimum threshold for the probability score to decide that the semantic shot prediction is confident enough.

Different values for the three parameters have been considered by running a grid search with the following ranges:

- **Minimum score:** from 0 to 1 in steps of 0.1.
- **Minimum number of elements:** from 1 to 5 in steps of 1.
- **Maximum radius:** from 0.1 to 1 in steps of 0.1.

For all possible combinations, the minimum number of elements and the maximum radius of clusters are used as input parameters to train the model. When detecting, the minimum score value is used at the decision stage.

For each combination, three cross-validation iterations have been run using the repeated random sub-sampling method. In each of the three trials, 80% of the data were randomly selected for training and the remaining 20% for test.

After all iterations, the following measures have been calculated out of the ones obtained from the addition of confusion matrices:

- **Precision and recall for each class:** provide a complete view of the performance for each class. They measure the exactness and completeness of the results, respectively.
- **F1 measure for each class:** the F1 measure combines the precision and recall values in a unified score, as described in Eq. 1.
- **Mean F1 measure:** the class-dependent F1 measures are averaged to obtain a single value that determines the optimal configuration of parameters.

$$F1 = 2\frac{precision \cdot recall}{precision + recall} \qquad (1)$$

Soccer. The best averaged F1 score has been obtained from the combination of clusters with 1 element, a cluster radius of 0.8 and a detection threshold of 0.7. Fig. 6 shows the accumulated confusion matrix after the three cross-validation trials and Fig. 7 a bar graph of the F1 score, precision and recall for every class.

Parliament. In the *parliament* case the averaged F1 score was maximized from the combination of clusters with at least 3 elements, a maximum radius of 0.8 and a detection threshold of 0.9. Fig. 8 shows the accumulated confusion matrix after the three cross-validation trials and Fig. 9 the F1 score, precision and recall for every class.

3   User Interface

The presented semantic shot detector has been installed on a web server which is remotely accessed from a browser. An intuitive Graphical User Interface (GUI)

		Automatic						
		Class1	Class2	Class3	Class4	Class5	Class6	Class7
	Class1	**652**	11	0	0	0	0	33
	Class2	4	**482**	0	0	0	0	36
Manual	Class3	1	0	**20**	0	0	0	3
	Class4	0	0	0	**94**	0	0	20
	Class5	0	0	0	0	**18**	0	6
	Class6	0	1	0	0	0	**21**	14
	Class7	461	3	0	0	0	0	**1204**

Fig. 6. Confusion matrix for Soccer

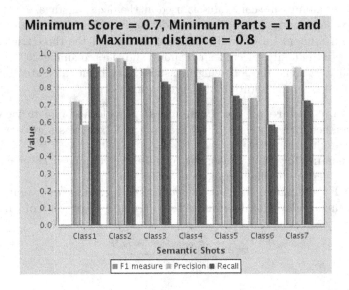

Fig. 7. Performance metrics for Soccer

		Automatic					
		Class1	Class2	Class3	Class4	Class5	Class6
	Class1	**93**	0	0	0	0	3
	Class2	0	**376**	0	0	0	2
Manual	Class3	0	0	**405**	0	0	3
	Class4	0	0	0	**75**	0	3
	Class5	0	0	0	0	**135**	3
	Class6	0	39	0	0	0	**465**

Fig. 8. Confusion matrix for Parliament

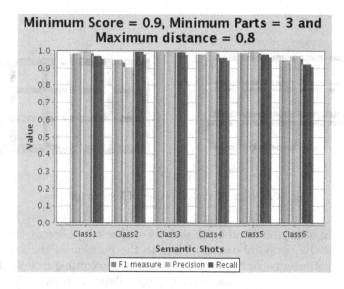

Fig. 9. Performance metrics for Parliament

has been developed to assist the archivists from the broadcasting company to label the keyframes with their semantic shot type. The GUI offers two basic functionalities: visualization of the suggestions for semantic shot labels, and tools for the edition and validation of the annotation. This section presents the developed tools and a proposed workflow.

3.1 Design

The automatic labels suggested by the system can be explored through the two different types of tabs shown in Fig. 10. The first layout displays the keyframes in the asset organized in rows, each of them for a semantic shot type. This is the default view after the user sends a video asset for its automatic annotation. The rows of the grid are dynamically filled as soon as new classification results arrive from the server. On the other hand, the second design shows groups of keyframes labelled with the same shot type. This option allows a fast overview of annotated keyframes belonging to the same class and allows a visual identification of any outlier.

In both cases the *drag and drop* mechanism has been chosen to be the primary interaction mode for fixing wrong detections. This mechanism can be used in three different situations: drag a keyframe from a panel (tab or row) and drop it to another, drag multiple keyframes from the same panel and drop them to another using the control key, or drag multiple keyframes from several panels and drop them to a another panel using the control key.

Once the keyframes are correctly labelled, the system awaits for a manual valida-tion from the user side. Two different methods have been implemented to confirm detections: validate all keyframes shown in the tab or validate all keyframes located in a row, all of them belonging to the same semantic shot type.

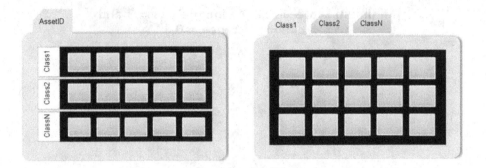

Fig. 10. Design of the two types of layouts for tabs

The sorting of keyframes in a tab can also follow two different criteria. The default one is the temporal one according to the position of the keyframes in the asset. An alternative option sorts the keyframes according to their detection score, a solution that may allow a faster page validation because higher score labels are supposed to be more confident that lower ones. Although a default minimum score has been learned for every domain, the interface also lets the user change this threshold so that all keyframes with a lower score will be placed in the *clutter* class panel. The implementation of this tool is possible because at the server side the detection engine does not apply any minimum threshold, so that all keyframes are assigned to a non-clutter class. The detection as *clutter* is performed at the client side according to this user-defined threshold.

3.2   Workflow

The semi-automatic annotation of an asset begins with the selection of the domain of interest (*soccer, parliament, ...*), which determines the models in the remote server that are to be used. The next step requires the introduction of the video asset ID to be annotated. At this point the interface will check with the content repository if the asset exists. If so, a preview of its keyframes will be shown to the user so that the ID can be confirmed, otherwise an error message will inform that the given ID is incorrect. Afterwards the IDs of the keyframes in the selected asset are sent to the annotation server, which will predict a semantic shot type for every image. The suggested labels are displayed on the *assets* tab and tools are provided to the user to edit or validate the results. Finally, when the labelling is satisfactory, it is validated and the annotation saved.

The interface has been developed using the technology provided by the Google Web Toolkit (GWT) version 2.2.0. The whole interface has been designed using strictly GWT panels and widgets, avoiding other GWT-based frameworks. A screenshot of the final result is shown in Fig. 11 and two video demos can be watched online[1,2].

---

[1] http://youtu.be/uURx9GoRJBg

[2] http://youtu.be/R7o0sCoe-Gs

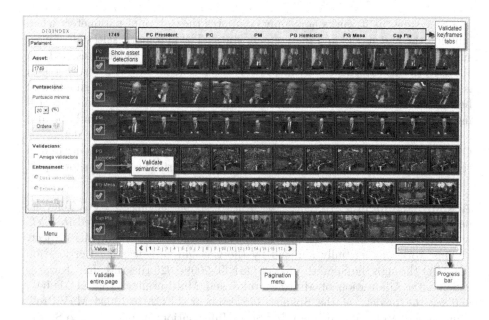

Fig. 11. Graphical User Interface

4 Conclusions

This paper has presented the development of a tool for the automatic detection of semantic shot types and its integration with a graphical user interface through a web service. The whole project has followed the requirements and guidelines collected from the team of archivists in a broadcast company (CCMA), who have defined the ontologies of semantic shot types and the design and workflow for the user interface.

A semantic shot detector has been developed using standardized MPEG-7 visual descriptors and SVM classifiers. This generic approach allows the system to be exploited in multiple domains, such as the two real-life examples of *soccer* and *parliament* defined by the archivists. The developed classifiers, originally designed for a binary detection, have been combined to define a more complex multiclass classifier. The obtained performance results show a high accuracy in the results, which allow considering the system for exploitation.

The developed GUI is easy to use but complete at the same time, providing all necessary information the archivists may need. The intuitive drag and drop mechanism naturally helps the user to edit the multiclass labels suggested by the system. The tools provided to work with the detections as sorting by score, changing the threshold or validating either the entire page or just a specific semantic shot, offer great flexibility. The resulting tool has been positively evaluated by the documentation department of the broadcast company, who have described it as "complete and easy to use".

The system, as it works now, only computes the optimum parameters when building the first model and they remain static. The automatic selection of classifier parameters is to be added in the future using the new validated results from the user. In this case, a strategy must be decided to define how new training data is to be used to updated the models. The interface and web service should also be further expanded to allow archivists define and train new domains according to their needs. Moreover, the detection of semantic shot types could be also used as a pre-filtering combined for more specific detectors (eg. face detectors, OCR...) so that, instead of assessing them to every single keyframe generated in the repository, they could be selectively applied to certain semantic shot types.

Acknowledgements. All images used in this paper belong to TVC, Televisió de Catalunya, and are copyright protected. They have been provided by TVC with the only goal of research under the framework of the BuscaMedia project.

This work was partially founded by the Catalan Broadcasting Corporation (CCMA) through the Spanish project CENIT-2009-1026 BuscaMedia:"Towards a Semantic Adaptation of Multinetworkd and Multiterminal Digital Media", and by the project of the Spanish Government TEC2010-18094 MuViPro:" Multicamera Video Processing using Scene Information: Applications to Sports Events, Visual Interaction and 3DTV".

References

1. Wang, M., Hua, X.-S., Song, Y., Lai, W., Dai, L.-R., Wang, R.-H.: An Efficient Automatic Video Shot Size Annotation Scheme. In: Cham, T.-J., Cai, J., Dorai, C., Rajan, D., Chua, T.-S., Chia, L.-T. (eds.) MMM 2007, Part I. LNCS, vol. 4351, pp. 649–658. Springer, Heidelberg (2006),
   http://dx.doi.org/10.1007/978-3-540-69423-6_63
2. Duan, L., Xu, M., Xu, C.-S., Jin, J.S.: A unified framework for semantic shot classification of sports video. IEEE Trans. on Multimedia 7(6), 1066–1083 (2005)
3. Bertini, M., Del Bimbo, A., Ioannidis, I., Stan, A., Bijk, E.: A flexible environment for multimedia management and publishing. In: 1st ACM International Conference on Multimedia Retrieval, art. 76. ACM, New York (2011),
   http://dx.doi.org/10.1145/1991996.1992072
4. Borth, D., Ulges, A., Breuel, T.M.: Lookapp: interactive construction of web-based concept detectors. In: 1st ACM International Conference on Multimedia Retrieval, art. 66. ACM, New York (2011), http://dx.doi.org/10.1145/1991996.1992062
5. Heyer, L.J., Kruglyak, S., Yooseph, S.: Exploring Expression Data: Identification and Analysis of Coexpressed Genes. Genome Research 9(11), 1106–1115 (1999)
6. Manjunath, B.S., Salembier, P., Sikora, T.: Introduction to MPEG-7. Wiley, West Sussex (2002)
7. Chang, C.-C., Lin, C.-J.: LIBSVM: a library for support vector machines. Trans. on Intelligent Systems and Technology 2(3) (2011),
   http://www.csie.ntu.edu.tw/~cjlin/libsvm

Labeling TV Stream Segments
with Conditional Random Fields

Emmanuelle Martienne[1], Vincent Claveau[2], and Patrick Gros[3]

[1] Université Rennes 2 - IRISA
Rennes, France
Emmanuelle.Martienne@univ-rennes2.fr
[2] CNRS - IRISA
Rennes, France
Vincent.Claveau@irisa.fr
[3] INRIA - Centre Rennes Bretagne Atlantique
Rennes, France
Patrick.Gros@inria.fr

Abstract. In this paper, we consider the issue of structuring large TV streams. More precisely, we focus on the labeling problem: once segments have been extracted from the stream, the problem is to automatically label them according to their type (eg. programs vs. commercial breaks). In the literature, several machine learning techniques have been proposed to solve this problem: Inductive Logic Programming, numeric classifiers like SVM or decision trees... In this paper, we assimilate the problem of labeling segments to the problem of labeling a sequence of data. We propose to use a very effective approach based on another classifier: the Conditional Random Fields (CRF), a tool which has proved useful to handle sequential data in other domains. We report different experiments, conducted on some manually and automatically segmented data, with different label granularities and different features to describe segments. We demonstrate that this approach is more robust than other classification methods, in particular when it uses the neighbouring context of a segment to find its type. Moreover, we highlight that the segmentation and the choice of features to describe segments are two crucial points in the labeling process.

1 Introduction

Now that TV is available in a digital form, it can be diffused using many channels and devices. Furthermore, the number of channels has drastically increased these last years, and huge amounts of video material are thus available. This is a good opportunity to develop new services and products, such as Catch-up TV services, TV on Demand, TV monitoring systems, or TV information retrieval systems. The point is that using directly TV streams, and not only programs or short extracts, remains difficult. While the stream is strongly structured from a human point of view, as a sequence of program separated by breaks, when observed at a lower level it is a continuous flow of pixels and audio frames with absolutely

E. Salerno, A.E. Çetin, and O. Salvetti (Eds.): MUSCLE 2011, LNCS 7252, pp. 183–194, 2012.

no apparent structure. The program guide, which is often diffused with the stream, is very inaccurate [6]. The breaks and even some short programs are not documented, and the start times provided are rough estimations that are not accurate enough for many applications. The problem is of course reinforced when dealing with continuous streams of several months. Any manual solution becomes very expensive and is not affordable for most applications. There is a need for automatic structuring tools. The structuring process can be split in different steps. A first one consists in segmenting the streams in entities that are assumed to correspond to semantic units like programs or breaks. During the second step, these segments are given labels to differentiate programs and breaks. The latter can be labeled in more precise categories like commercials, jingles, trailers, etc. The labeling of program segments can be achieved by aligning the stream with a program guide. The programs can then be decomposed in smaller sequences, e.g. the different topics of a news report can be segmented. In this paper, we focus on the labeling process, and more precisely, on classification based labeling. We assume that some segments have been extracted using a manual or a state-of-the-art method [4, 7, 16] and we try to classify the segments into programs and breaks, but also into different types of breaks. Some methods have already been proposed to solve such a problem: symbolic classifiers such as Inductive Logic Programming but also numeric ones like Markov models or SVM (see section 2). Their main drawback is the lack of robustness since they have been tested on very structured and stable streams, which is rarely the case. In this paper, we propose a more robust approach to the task of segments labeling. We assume that the problem of finding the type of a segment in a stream can be assimilated to the problem of finding the label that is associated with a particular element in a sequence of data. We also make the hypothesis that segments in a stream appear according to some repetitions schemes, and that the type of a segment is related to the types of its neighbouring segments. Therefore, we investigate the use of another tool that seems particularly adapted to the task of segments labeling: Conditional Random Fields (CRF). Indeed, CRF aim at inducing the sequence of labels that is associated with a sequence of observations. Moreover, during its induction process, CRF take into account previous and next observations and labels to find the label of a current observation. We compare CRF to other classification methods proposed in the state-of-the-art and show that CRF are more efficient. The remainder of this paper is organized as follows. Section 2 is a short survey of existing TV streams structuring methods. Section 3 presents the datasets that were used for the experiments described in the paper. Section 4 details the approach conducted to evaluate CRF capability to label TV segments. Sections 5 and 6 contain the results of different experiments performed according to this approach, and finally section 7 concludes and discusses future work.

2   Related Works

There are two main ways to tackle this TV streams structuring problem. Poli et al. [9] are the only ones to our knowledge who have developed a top-down approach. They make the assumption that the stream structure is very stable over the years and they try to learn it from a very large database. Afterwards, they use this learned structure to predict program boundaries and the sequence of breaks and programs. In a final step, they process the stream at predicted instants in order to refine the boundaries. On the one hand, such a method reduces enormously the part of the streams that has to be processed. This insures good performance and removes lots of errors. On the other hand, it requires a huge learning set: 4 years of manually annotated data in their experiments. Technically speaking, the model they learn is a temporized Markov model: the observation probabilities depend on the time and of the day of the week. The model performs well, but it cannot adapt to special events like holidays, sport events (e.g. Tour de France or Olympics) which cause major changes to the stream structure. Even more data would be necessary to learn such events. Bottom-up approaches are the most frequent ones. They try to infer the structure by studying the stream itself. Many approaches were designed to detect commercials or jingles, based on audio, visual or audio-visual features, but they do not address the structuring problem itself. Among the methods which address this problem, Naturel's [7] is probably one of the first in this category. It relies on the manual annotation of one day of stream. Afterwards, it segments the stream and detects repetitions between this day of reference and the subsequent segments of the stream. Some post-processing rules are applied in order to label also unrepeated segments. These new segments are added to the reference set in order to update it. This data set used in Naturel's paper was unfortunately too small (three weeks) to verify that this updating technique is robust enough in order to maintain the quality of the results for long streams. Manual annotation is a tedious task that cannot be done too often. In order to circumvent this annotation problem, Manson et al. [4] proposed to replace the propagation technique by a learning one. They train a symbolic machine learning system based on ILP (Inductive Logic Programming) that labels the segments according to their local context. Even more manual annotation is required to learn the structural information (three weeks), but the annotated part of the stream does not need to be contiguous to the unknown part. Furthermore, it is not the content of the annotated part, which is further used, but its structure that is captured by the learning system. As Poli showed it, the validity of such information should be much longer. Unfortunately, Manson's system was only tested with data in the 6:00 PM - 12:00 PM time slice that is known to be the most structured and stable one. Its robustness to less structured streams is still to be demonstrated. The present paper proposes the use of another kind of classifier in order to gain this missing robustness.

3   Data

The TV stream we deal with is a three weeks long broadcast of a French channel. Two datasets have been extracted from this stream : the first one, that will be refered as *manually-segmented dataset*, is the result of a manual segmentation of the stream. This dataset will allow us to evaluate the labeling process only, without any influence from an automatic segmentation tool. The second dataset, that will be called *auto-segmented dataset*, is the result of an automatic segmentation method based on the detection of repeated segments [1]. This dataset will allow us to evaluate our labeling approach in the context of an automatic segmentation. In the two datasets, segments are identified through their first and last image in the stream. Each segment has a label indicating whether it is a program or a break, and for breaks, four additional labels have been used, namely *commercial, trailer, sponsoring* and *jingle*. In addition, each segment is described with the date of its broadcast. Table 1 shows a sample of the data: each line describes a segment; segments appear in the chronological order. In the Label column, value 1 stands for a program and value 2 for a break. The Ad. label column provides the fine-grain classification for breaks: 1: commercial, 2: jingle, 3: sponsoring and 4: trailer. Segments labeled as programs always have the 1 value in the Ad. label column.

Table 1. Sample of the data

Date	Label	Ad. label	First image	Last image	Title
09052005	1	1	1	27258	Seg_{54}
09052005	2	4	27259	27563	Seg_{55}
09052005	2	3	27575	27764	Seg_{56}
09052005	2	2	27767	27834	Seg_{57}
09052005	2	4	27835	28758	Seg_{58}

Table 2 shows, for each dataset, the total number of segments and the distribution of these segments over the different types (program, commercial, etc). Note that the auto-segmented dataset contains much more segments than the manually-segmented one. Indeed, the use of an automatic segmentation method led to an over-segmentation of the initial stream.

Table 2. Distribution of the different types of segments over the two datasets

Dataset Type	manually-segmented	auto-segmented
program	1506	22557
trailer	1290	4075
commercial	1050	18089
sponsoring	1714	2201
jingle	2031	1622
total	7591	48544

These datasets were used as a basis for the experiments reported in this paper. Two types of features have been extracted to describe the segments, namely robust features and content-based features.

3.1   Robust Features

Robust features can be computed very efficiently, and are expected to be robust since they do not depend on the quality of the image or sound signal of the stream. In our case, robust features include duration as well as two properties related to the broadcast such as moment of the week (week day, week-end or holiday) and period of the day (night, morning, noon, etc).

Table 3. Sequence of segments described with robust features

Title	Moment of the week	Period of the day	Duration	Label
Seg_{33}	weekday	prime	$]0,10s[$	commercial
Seg_{34}	weekday	prime	$[30s,1m[$	jingle
Seg_{35}	weekday	prime	$[5m,+[$	program
Seg_{36}	weekday	postprime	$]0,10s[$	jingle
Seg_{37}	weekday	postprime	$]0,10s[$	trailer

3.2   Content-Based Features

Content-based features are more related to the properties of the segments within the TV stream. They can be divided into two parts. On the one hand, *global features* include the number of occurrences of a segment, the number of different calendar days it was broadcasted, the number of different days of the week (monday to sunday) it was broadcasted[1] and its duration. On the other hand, *local features* are extracted from two different sources of information: the first one is the presence of a separation[2] before and/or after a segment. The second one is the neighbouring segments of a current segment. Let W_b and W_a be respectively the temporal windows before and after the current segment. The features that are issued from these temporal windows are:

- the numbers of neighbouring segments in W_b and W_a;
- the total numbers of times the neighbouring segments of W_b and W_a are repeated in the stream;
- the minimal, maximal and average numbers of times the neighbouring segments of W_b and W_a are repeated in the stream.

In total, 15 content-based features are used to describe segments.

4   The Labeling Approach Using CRF

4.1   Conditional Random Fields

Conditional Random Fields [3] are a supervised learning framework whose main purpose is the annotation of data. Given an *observation* x, CRF aim at inducing

---

[1] For instance, for a segment broadcasted 10 consecutive tuesdays, the number of different days of the week is 1 whereas the number of different calendar days is 10.

[2] A separation is the simultaneous occurrence of monochrome frames and silence that happens between commercials in France (due to legal regulations).

its corresponding *annotation* y, from a set of already annotated examples, i.e. a set of pairs (x, y). x and y can be of different natures. For instance, CRF have been widely used in natural language processing to annotate XML trees [2] or syntactic trees [5]. In computational biology and computational linguistics, they have been applied to various problems such as part-of-speech (POS) tagging, text analyzing [8, 11] and RNA secondary structure analyzing [13]. CRF have also been useful in the multimedia field to solve some detection problems [15, 14, 10]. Recently, CRF were successfully applied to the labeling of sequential data, i.e. to induce the sequence of labels $y = (y_1, ..., y_n)$ that are associated with a sequence of observations $x = (x_1, ..., x_n)$. CRF are undirected graphical models (see figure 1). A model produced with CRF is conditional in the sense that it defines the probabilities of the possible sequences of labels y, given a sequence of observations x. Such a model is further used to label a new sequence of observations x', by selecting the sequence of labels y' that maximizes the conditional probability $P(y'|x')$. The main advantages of conditional models like CRF are twofold: (1) contrary to generative models like Hidden Markov Models (HMM), they do not need the enumeration of all the possible sequences of observations, and (2) the probabilities of labels can depend on past, current and further observations. The CRF framework seems particularly well suited to handle our own sequential data. Indeed, CRF will allow us to build a model where the label of a segment will depend on the labels of its neighbouring segments.

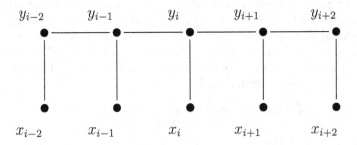

Fig. 1. Graphical structure of CRF for sequences

4.2   Labeling TV Segments with CRF

A television stream can be viewed as a sequence of video segments, where segments features are observations and segments types (i.e. programs, commercials, trailers, etc) are labels. Therefore, the problem of labeling TV segments can be compared to the problem of labeling sequential data that is addressed by CRF. The goal of this paper is to evaluate the relevance of the CRF framework in the context of video segments labeling. The question we ask is: can CRF be efficient to predict segments labels given segments features? The process that is conducted to evaluate CRF in this framework is a standard process for the evaluation of learning methods. A first stream, called *the training stream*, is used by CRF to build the probabilistic model. Then, this model is tested on a second

stream (different from the learning stream), called *the test stream*. At the end
of the test stream labeling, the comparison of the real and the predicted labels
of the segments allows for calculating some measures, like *precision* and *recall*,
that evaluate the relevance of the learned model to automatically label a new
stream. Both streams are extracted from the datasets previously described (see
section 3). Segments in the streams are described with the same features (robust
or content-based).

5   Experiments on Manually-Segmented Dataset

The first experiments have been performed using the manually-segmented dataset
and describing segments with robust features. The goal of these experiments is
to evaluate the quality of the labeling with CRF, in the context of a perfect
segmentation of the stream where segments are described with features that are
easy to compute. The first two weeks of the TV stream have been used as the
training stream, and the third remaining week as the test stream. In order to vary
the label grain, two experiments have been conducted: in the first experiment,
segments are labeled as programs or breaks. In the second one, a distinction is
made between the different types of breaks. Thus, five labels are used : program,
commercial, trailer, jingle and sponsoring. The CRF open source implementation
we used to build the probabilistic model is *CRF++*, developed by Taku Kudo[3].
In CRF++, *feature templates* must be provided as parameters of the system.
These feature templates indicate, given the current element that is examined
in the sequence, the set of observation features that can be used to compute
probabilities. The feature templates used in all the experiments reported in this
paper are:

- the features and label of the current segment;
- the features of the four previous segments in the stream;
- the features of the four next segments in the stream.

The highlight of the CRF method is that it learns by taking into account the
sequentiality of the data. To prove that sequentiality is a crucial point in the
process of segments labeling, we need to compare the results of CRF with those
of other classification methods that do not make use of this sequentiality. Consequently, the same experiments with different label granularity have also been performed with other classification methods by using the same manually-segmented
dataset, under the same learning context. The following tables 4 and 5 show the
results of the whole experiments. Only the results of the three best classification
methods are given.

We can first notice that the CRF results are higher than those of other classification methods, especially in the case of a multiple labeling that separates the
different types of breaks. The handling of the sequentiality of the data seems really important in the labeling process. In the case of multiple labeling, precisions

---

[3] http://crfpp.sourceforge.net/

Table 4. Results of CRF and other classification methods - Program vs Break label

Method	Label	Precision	Recall	F-measure
CRF	Break	95.49%	97.80%	96.63%
	Program	90.61%	82.09%	86.14%
KNN	Break	91.92%	95.28%	93.57%
	Program	78.72%	67.55%	72.71%
C4.5	Break	94.63%	88.69%	91.56%
	Program	64.76%	80.50%	71.78%
Classification tree	Break	94.21%	89.33%	91.71%
	Program	65.58%	78.72%	71.55%

and recalls of the CRF model have quite high values (from 78.27% to 88.43% for the precision and from 77.09% to 88.12% for the recall), that vary from a label to another. However, it seems more difficult for the model to distinguish between the different types of break segments since precision and recall are higher in the case of a Program vs Break labeling. Indeed, in the confusion matrix (see table 6) we can see that the model often confuses between jingles and sponsorings: 125 segments that are really jingles are labeled as sponsorings. Conversely, 151 segments that are really sponsorings are labeled as jingles. We observe the same confusion between commercials and programs. This can be explained by the fact that the very simple and robust features used to describe the segments are not sufficient to distinguish between break segments of different types that still have similar properties.

6   Experiments on Auto-segmented Dataset

Experiments on the manually-segmented dataset have shown the relevance of CRF to label segments when the segmentation is quite perfect. However, segmenting TV streams manually is not a realistic approach. The goal of these experiments is to evaluate CRF on data that result from an automatic segmentation. The dataset that is used is the auto-segmented dataset, where segments are described with robust features. The experiments performed are the same as those described in section 5. The following tables 7 and 8 show the results of CRF and other classification methods.

In section 3, we pointed out the over-segmentation of the TV stream that results from the use of an automatic segmentation method. This over segmentation has a great impact on the labeling. Indeed, CRF and other classification methods are less efficient at separating programs from breaks. CRF still has the best results, but it seems that the number of consecutive segments sharing the same features is so high that it becomes very difficult to identify their true types. The results really get worse for all the methods in the case of a multiple labeling: jingles and sponsorings are almost not detected, precisions and recalls are very low on most types of breaks.

The previous experiments and those described in section 5 show that the robust features are sufficient to discriminate the different types of segments, only

Table 5. Results of CRF and other classification methods - Multiple labels

Method	Label	Precision	Recall	F-measure
CRF	Program	88.43%	88.12%	88.27%
	Trailer	88.33%	87.40%	87.86%
	Jingle	78.27%	82.85%	80.49%
	Commercial	84.51%	81.23%	82.84%
	Sponsoring	79.89%	77.09%	78.47%
KNN (10NN)	Program	78.76%	67.73%	72.83%
	Trailer	93.27%	72.70%	81.71%
	Jingle	50.08%	86.57%	63.45%
	Commercial	50.36%	66.99%	57.50%
	Sponsoring	59.22%	20.35%	30.29%
Naïve Bayes	Program	67.79%	64.18%	65.94%
	Trailer	93.02%	83.99%	88.27%
	Jingle	50.60%	89.10%	64.54%
	Commercial	44.03%	45.31%	44.66%
	Sponsoring	63.60%	19.54%	29.90%
SVM	Program	65.58%	78.72%	71.55%
	Trailer	93.02%	83.99%	88.27%
	Jingle	50.60%	89.10%	64.54%
	Commercial	44.57%	25.24%	32.23%
	Sponsoring	63.60%	19.54%	29.90%

Table 6. Confusion matrix of our CRF model for each segment label

		Real				
		Jingle	Spons.	Trailer	Com.	Prog.
Predicted	Jingle	623	151	21	0	1
	Spons.	125	572	14	1	4
	Trailer	2	13	333	6	23
	Com.	0	2	5	251	39
	Prog.	2	4	8	51	497

Table 7. Results of CRF and other classification methods - Program vs Break label

Method	Label	Precision	Recall	F-measure
CRF	Break	64.97%	90.34%	75.58%
	Program	81.63%	46.83%	59.52%
SVM	Break	58.65%	72.41%	64.81%
	Program	59.52%	44.28%	50.78%
CN2	Break	59.05%	62.03%	60.50%
	Program	56.14%	53.04%	54.55%
Classification tree	Break	59.09%	61.99%	60.51%
	Program	56.16%	53.16%	54.62%

Table 8. Results of CRF and other classification methods - Multiple labels

Method	Label	Precision	Recall	F-measure
CRF	Program	67.17%	64.61%	65.87%
	Trailer	10.17%	1.23%	2.19%
	Jingle	0%	0%	0%
	Commercial	49.26%	74.37%	59.27%
	Sponsoring	20.45%	0.94%	1.80%
Classification tree	Program	51.81%	69.64%	59.42%
	Trailer	25.93%	0.48%	0.94%
	Jingle	0%	0%	0%
	Commercial	39.94%	40.72%	40.33%
	Sponsoring	0%	0%	0%
CN2	Program	51.78%	69.78%	59.45%
	Trailer	26.47%	0.62%	1.21%
	Jingle	0%	0%	0%
	Commercial	39.87%	40.41%	40.14%
	Sponsoring	0%	0%	0%
C4.5	Program	51.80%	72.02%	60.26%
	Trailer	0%	0%	0%
	Jingle	0%	0%	0%
	Commercial	39.96%	38.40%	39.16%
	Sponsoring	0%	0%	0%

Table 9. Results of CRF and other classification methods - Program vs Break label

Method	Label	Precision	Recall	F-measure
CRF	Break	93.72%	94.76%	94.24%
	Program	93.26%	91.95%	92.60%
SVM	Break	95.07%	91.92%	93.47%
	Program	90.17%	93.96%	92.03%
CN2	Break	94.78%	91.97%	93.35%
	Program	90.19%	93.58%	91.85%
Naïve Bayes	Break	91.52%	94.87%	93.16%
	Program	93.18%	88.86%	90.97%

when the segmentation is manual. They are no more relevant when the TV
stream is automatically segmented. Thus, the time saved with the use of easy to
compute features is wasted by the need to manually segment the stream. The
last experiments described below aim at checking if other features, in our case
the content-based features, are more relevant in the case of an automatically segmented stream. These experiments have been conducted on the auto-segmented
dataset where the segments are described with content-based features (see section 3.2) and labeled as programs or breaks. To take into account the large
volume of the dataset, the first 30% of the segments have been used as the training stream and the remaining 70% as the test stream. The following table 9
shows the results.

We can see that the use of content-based features really increases the performances of all the classifiers. Indeed, the results are better even with fewer training data.

7   Conclusion and Future Work

In this paper, we addressed the problem of labeling TV segments using Conditional Random Fields, a classification method whose specificity is to handle sequential data. CRF is a particularly well-suited approach for the task of labeling TV segments. The experiments conducted have proved the robustness of the method to separate program segments and break segments, but also to distinguish between different types of break segments when the segments are manually or automatically segmented. Results of CRF exceed those of other classification methods. Moreover, these experiments have also highlighted the fact that basic features are limited to detect programs and breaks, especially when the segmentation is performed automatically, resulting in an over-segmented stream. In that case, content-based features must be used to separate the different kinds of segments. The main limit of CRF is that it is a supervised learning method, i.e. a method that learns from a set of already labeled data. Manually labeling a part of the segments is an unrealistic and tedious task. The goal of our future work will be to reduce the need for training data. In particular, we'll study the possibility to use CRF in a non-supervised learning context, by inspiring from similar work that have been conducted with other learning methods [12].

References

[1] Ibrahim, Z.A.A., Gros, P.: Tv stream structuring. ISRN Signal Processing (2011)
[2] Jousse, F.: Transformation d'arbres XML avec des modèles probabilistes pour l'annotation. PhD thesis, University of Lille III, France (2007)
[3] Lafferty, J., McCallum, A., Pereira, F.: Conditional random fields: Probabilistic models for segmenting and labeling sequence data. In: Proc. of the Int. Conf. on Machine Learning (ICML), pp. 282–289 (July 2001)
[4] Manson, G., Berrani, S.-A.: An inductive logic programming-based approach for TV stream segment classification. In: Proc. of the IEEE Int. Symp. on Multimedia (December 2008)
[5] Balvet, A., Laurence, G., Rozenknop, A., Moreau, E., Tellier, I., Poibeau, T.: Annotation fonctionnelle de corpus arborés avec des champs aléatoires conditionnels. In: TALN
[6] Naturel, X.: Structuration automatique de flux vidéos de télévision. PhD thesis, University of Rennes 1, France (2007)
[7] Naturel, X., Gravier, G., Gros, P.: Fast Structuring of Large Television Streams Using Program Guides. In: Marchand-Maillet, S., Bruno, E., Nürnberger, A., Detyniecki, M. (eds.) AMR 2006. LNCS, vol. 4398, pp. 222–231. Springer, Heidelberg (2007)
[8] Pinto, D., McCallum, A., Wei, X., Croft, W.: Table extraction using Conditional Random Fields. In: Proc. of the ACM SIGIR, pp. 235–242 (July 2003)

[9] Poli, J.-P.: An automatic television stream structuring system for television archives holders. Multimedia systems 14(5), 255–275 (2008)

[10] Quattoni, A., Collins, M., Darrell, T.: Conditional random fields for object recognition. In: Neural Information Processing Systems (NIPS) (December 2004)

[11] Sha, F., Pereira, F.: Shallow parsing with conditional random fields. In: Proc. of Human Language Technology - North American Chapter of the Association for Computational Linguistics (May 2003)

[12] Shi, T., Horvath, S.: Unsupervised learning with random forest predictors. Journal of Computational and Graphical Statistics (2006)

[13] Tabei, Y., Asai, K.: A local multiple alignment method for detection of non coding RNA sequences. Bioinformatics 25, 1498–1505 (2009)

[14] Weinman, J., Hanson, A., McCallum, A.: Sign detection in natural images with Conditional Random Fields. In: Proc. of the IEEE International Workshop on Machine Learning for Signal Processing (September 2004)

[15] Yuan, J., Li, J., Zhang, B.: Gradual transition detection with conditional random fields. In: Proc. of the 15th International Conference on Multimedia, pp. 277–280. ACM (September 2007)

[16] Zeng, Z., Zhang, S., Zheng, H., Yang, W.: Program segmentation in a television stream using acoustic cues. In: Proc. of the International Conference on Audio, Language and Image Processing, pp. 748–752 (July 2008)

Foreground Objects Segmentation
for Moving Camera Scenarios Based on SCGMM

Jaime Gallego, Montse Pardàs, and Montse Solano*

Technical University of Catalonia (UPC),
Department of Signal Theory and Communications,
st. Jordi Girona 1-3, Office D5-120, zip: 08034, Barcelona, Spain
{jgallego,montse}@gps.tsc.upc.edu

Abstract. In this paper we present a new system for segmenting non-rigid objects in moving camera sequences for indoor and outdoor scenarios that achieves a correct object segmentation via global MAP-MRF framework formulation for the foreground and background classification task. Our proposal, suitable for video indexation applications, receives as an input an initial segmentation of the object to segment and it consists of two region-based parametric probabilistic models to model the spatial (x,y) and color (r,g,b) domains of the foreground and background classes. Both classes rival each other in modeling the regions that appear within a dynamic region of interest that includes the foreground object to segment and also, the background regions that surrounds the object. The results presented in the paper show the correctness of the object segmentation, reducing false positive and false negative detections originated by the new background regions that appear near the region of the object.

Keywords: Object segmentation, SCGMM, moving camera sequences, video indexation.

1 Introduction

Objects segmentation and tracking in moving camera scenarios is of main interest on several high level computer vision applications like human behavior analysis or video sequence indexation among others, where a specific segmentation of the object, previously determined by the user, is needed. This kind of scenarios are the most common in video recordings, but present a special challenge for objects segmentation due to the presence of relative motion concerning the camera observer point and the foreground object to segment, which causes a non-stationary background along the sequence. Therefore, this scenario differs from fixed camera ones, where an exact background can be learned at a pixelwise level [1,2] and fixed camera with constrained motion scenarios, typical of

---

* This work has been partially supported by the Spanish Ministerio de Ciencia e Innovación, under project TEC2010-18094, and the UPC Ph.D. program.

E. Salerno, A.E. Çetin, and O. Salvetti (Eds.): MUSCLE 2011, LNCS 7252, pp. 195–206, 2012.

surveillance cameras with a programmed camera path, which can be considered as a static mosaic from the dynamic scenes [3]. Instead, moving camera scenarios present a more difficult framework due to the impossibility of applying well known pixel-wise techniques for computing the background subtraction, and it has led to the publication of several new proposals that addresses this topic in the last few years. [4] presents a review of the most recent background segmentation systems.

The different techniques proposed in previous works can be grouped into three classes:

-Techniques based on camera motion estimation. These methods compute camera motion and, after its compensation, they apply an algorithm defined for fixed camera. [5] uses frame differencing and active contour models to compute the motion estimation. In [6], the authors apply background subtraction using the background obtained through mosaicing numerous frames with warping transforms, while [7] proposes a multi-layer homography to rectify the frames and compute pixel-wise background subtraction based on Gaussian Mixture Model.

-Methods based on motion segmentation. In these methods the objects are mainly segmented by analyzing the image motion on consecutive frames. [8] proposes to use image features to find the optic flow and a simple representation of the object shape. [9] proposes a semi-automatic segmentation system where, after a manually initialization of the object to segment, a motion-based segmentation is obtained through region growing algorithm. In [10] an approach based on a color segmentation followed by a region-merging on motion through Markov Random Fields is proposed.

-Based on probabilistic models: the objects to segment are modeled using probabilistic models that are used to classify the pixels belonging to the object. [11] proposes a non parametric method to approximate, in each frame, a p.d.f. of the objects bitmap, estimating the maximum a posteriori bitmap and marginalizing the p.d.f over all possible motions per pixel.

The main weakness of the systems based on motion estimation is the difficulty to estimate the object or camera motion correctly and the impossibility of subtracting the background when dynamic regions are present, which produces many false positive detections. On the other hand, proposals based on using foreground object probabilistic models present a more robust segmentation, but can lead to segmentation errors when the close background presents similar regions to the object.

In this paper we propose a new technique for object segmentation in moving camera scenarios that deals with the last group of segmentation methods based on probabilistic models. We propose to use a region-based parametric probabilistic model, the Spatial Color Gaussian Mixture Model (SCGMM) to model not only the foreground object to segment, but also the close-background regions that appear surrounding the object, allowing, in this manner, a more robust classification of the pixels into foreground and background classes. The use of this novel technique achieves a correct segmentation of the foreground object via global MAP-MRF framework for the foreground (fg) and background (bg) classification task.

The remainder of the paper is organized as follows. Section 2 describes the scene motion model that we propose to define the context of our proposal. Section 3 shows the proposed object segmentation system. Finally, some results and conclusions are presented in Section 4 and Section 5 respectively.

2   Motion Model System

The main strategies of the state of the art to achieve the segmentation of a certain object in a moving camera scenario, focus on analyzing two main factors: the scene motion between frames and the object characteristic features. These proposals are based on the principle that this kind of sequences present two different motions corresponding to the camera and to the object to segment.

We propose a new framework to solve the segmentation problem. Consider a moving camera sequence, where the camera performs some movements of translation, rotation and zoom and the object to segment is also moving inside the scene, changing also its orientation and making some rotations.

We will consider that the camera translation and rotation effects, together with the object orientation and translation changes are equivalent to consider a background motion behind the object to segment.

Therefore, using a dynamic region of interest, centered in the object detection obtained in the previous image, we will consider that the background is a plane located behind the object to segment, which suffers some spatial modifications along the sequence and where new background regions appear in the limits of the image (usually due to camera displacements).

To perform the segmentation we will use two probabilistic models: One to model the foreground object to segment, and another to model the background that is surrounding the object, with the objective that the background model assumes the new background regions that appear close to the object, achieving a robust classification process of the pixels among the two classes. Both models must also be flexible to assume possible camera zoom and object rotations that occur along the sequence.

3   Proposal

The scene model that we use, allows several spatial transformations: camera zoom, foreground object rotations and background rotation and translation. We propose a versatile segmentation system that allows us to overcome all these situations, which consists of two separated parametric models to model the foreground object to segment and the close background that envelopes the object. For this purpose, we will use the Spatial Color Gaussian Mixture models (SCGMM) [12], which have proved to work efficiently in most considered scenarios [14,15]. The system works as follows:

At the beginning, the system needs an input mask of the object that we want to segment. This region mask can be obtained via manually segmentation or using any segmentation tool, and it is used to:

- Define the dynamic Region of Interest of the object, defined as the bounding box that encloses the object with a percentage of close background.
- Initialize the foreground and the close background SCGMM that appear inside the already defined objects' ROI.

For each frame of the sequence, there is a three steps process: Classification of each pixel inside the bounding box according to the foreground and background models defined from the previous frame, updating of each model using the results obtained from the classification step and redefinition of the ROI according to the resultant foreground object segmentation. Figure 1 shows the overall work flow diagram. The details of this segmentation system will be explained in the following subsections.

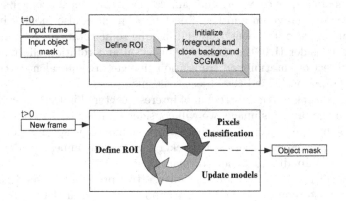

Fig. 1. Work flow of the proposed system

3.1   Dynamic Region of Interest

According to the motion model that we have defined, we propose to make a local foreground object segmentation. We define the background model within a dynamic bounding box surrounding the foreground object. This neighborhood is defined according to some constraints of computational cost, and accuracy in the background modeling.

The bounding box has to present a certain size that allows the background model to achieve a correct close background representation in all the boundaries of the object, allowing all possible movements of the object to segment, but it has to be small enough to allow a reduced computational cost when updating models or calculating pixel probabilities. The model used has to be flexible enough to incorporate new parts of the background that appear around the object as the camera or the object move along the scene.

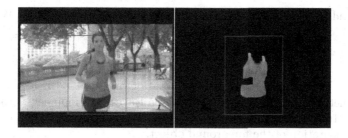

Fig. 2. Dynamic Region of interest over the initialization mask

Thus, the bounding box will be centered at the geometric center of the object, with the limits of the object to segment plus an offset d that we define as a percentage of the largest axis of the ellipse that envelopes the object. 20% yields correct results in most considered scenarios. Figure 2 show a graphical example of this bounding box.

3.2   Probabilistic Models

A good segmentation of foreground objects can be achieved if a probabilistic model for the foreground and also for the close background are constructed. Hence, we classify the pixels in foreground (fg) and background (bg) classes. Since in this kind of sequences the foreground and background are constantly moving and changing, an accurate model at a pixel level is difficult to build and update. For this reason, we use a region based Spatial Color Gaussian Mixture Model (SCGMM), as in [12,14,15], because foreground objects and background regions are better characterized by color and position, and GMM is a parametric model that describes accurately multi-modal probability.

Thus, the foreground and background pixels are represented in a five dimensional space. The feature vector for pixel i, $zi \in \mathbb{R}^5$, is a joint domain-range representation, where the space of the image lattice is the domain, (x, y), and the color space, (r, g, b), is the range [13]. The likelihood of pixel i is then,

$$P(z_i|l) = \sum_{k=1}^{K_l} \omega_k G_l(z_i, \mu_k, \Sigma_k)$$

$$= \sum_{k=1}^{K_l} \omega_k \frac{1}{(2\pi)^{5/2}|\Sigma_k|^{1/2}} e^{-\frac{1}{2}(z_i-\mu_k)^T \Sigma_k^{-1}(z_i-\mu_k)}$$

where l stands for each class: $l = \{\text{fg}, \text{bg}\}$, ω_k is the mixture coefficient, μ_k and Σ_k are, respectively, the mean and covariance matrix of the k-th Gaussian distribution, $|\Sigma_k|$ is the determinant of matrix Σ_k. It is commonly assumed that the spatial and color components of the SCGMM are decoupled, i.e., the covariance matrix of each Gaussian component takes the block diagonal form,

$$\Sigma_k = \begin{pmatrix} \Sigma_{k,s} & 0 \\ 0 & \Sigma_{k,c} \end{pmatrix}$$

where s and c stand for the spatial and color features respectively. With such decomposition, each foreground Gaussian component has the following factorized form:

$$G_l(z_i, \mu_k, \Sigma_k) = G(x_i, \mu_{k,s}, \Sigma_{k,s})G(v_i, \mu_{k,c}, \Sigma_{k,c}), \qquad (1)$$

where $x_i \in \mathbb{R}^2$ is the pixel's spatial information and $v_i \in \mathbb{R}^3$ is its color value. The parameter estimation can be reached via Bayes' development, with the EM algorithm [16]. For this estimation an initialization frame is needed, containing a first segmentation of the foreground object.

Initialization. Once we have defined the Bounding box where the foreground and background models will work, the initialization of both models is done according to the object mask that is required as an input.

The number of Gaussians that will compound each model should be slightly higher than the number of color-spatial regions of the foreground and background regions that appear within the ROI, to ensure that both classes are correctly modeled with at least one Gaussian per region. There are several manners to obtain this number of regions. In our case, we choose to analyze the RGB-histogram in the following way: Once the foreground and background histograms are calculated, the number of bins used to define them are examined to detect the N first bins with higher probability which gather together the 70% of the color appearance probability. In each class, for each one of these bins, a Gaussian will be added to the model. In this way, we obtain a model with the correct number of gaussians to represent the foreground object and the close background regions.

Once the number of Gaussians of each model is defined, we propose a fast two-steps initialization process that consists in:

- First, place the Gaussian distributions of the foreground and background models uniformly over the spatial region that corresponds to each model.
  We initialize the spatial and color domain of the Gaussians with the values of the pixels that are located within the region assigned to each Gaussian. Figure 3 displays a graphical and self-explicative example.
- Next, for each class, we use the Expectation Maximization (EM) algorithm ([16]) in the overall five dimensional domain with the color-spatial information obtained from all the pixels belonging to the class we are analyzing, and located inside the ROI. This algorithm helps us to adjust the parameters of each Gaussian Mixture Model in the color and spatial domain, $\mu_{c,s}$ and $\Sigma_{c,s}$ of each model obtaining iteratively a maximization of the likelihood. Thanks to the spatially uniform distribution of the Gaussians, the initialization requires a few EM iterations to achieve the convergence of the algorithm and therefore, a correct representation of the foreground and background regions. A fix number of iterations equal to 3 yields correct results. Figure 3 shows the resultant initialization of the Gaussians in the spatial domain.

Updating. We assume a scene with moving background, moving foreground object as well as possible zoom effects of the camera, where new color-spatial

Fig. 3. Initialization process. From left to right: spatially uniform distribution of the Gaussians, Foreground Gaussians after EM iterations and Background Gaussians after EM iterations. The spatial domain representation of the foreground Gaussians is in red color, background Gaussians are in green color.

regions of background and foreground classes inside the Region of Analysis appear in each frame. Thus, the spatial components of each Gaussian Mixture and also, the color ones, need to be updated after the classification in foreground and background of each frame. The complete updating of both classes in the spatial and color domains could lead to False Positives error propagation if the foreground regions present similar colors to the background.

Thus, we propose a two-steps updating for each model that allows a correct spatial domain updating and a conditional color updating which introduces new color regions to the model depending on the degree of similarity between the foreground and background models. The two steps updating is as follows:

Spatial Domain Updating: The pixels classified as foreground and background, form a mask that is used for the updating of each class respectively. In this step, only the spatial components of the Gaussian Mixture are updated ([14]). As it is proposed in [17], we assign each pixel to the Gaussian k that maximizes:

$$P(k|z_i, l) = \frac{\omega_k G_l(z_i, \mu_k, \sigma_k)}{\sum_k \omega_k G_l(z_i, \mu_k, \sigma_k)} \tag{2}$$

the denominator is the same for all the classes and can be disregarded.

Once each pixel has been assigned to a Gaussian, the spatial mean and covariance matrix of each Gaussian are updated with the spatial mean and variances of the region that it is modeling.

Also, in order to achieve a better adaptation of the model into the shape, we propose a Gaussian split criterion according to the spatial size of the Gaussian. The Gaussians that accomplish the following expression are split into two smaller Gaussians in the direction of the eigenvector associated to the largest eigenvalue, $\lambda_{\max}$: $\lambda_{\max} > \chi$, where χ is a size threshold. In our tests, $\chi = \max(\text{object}_{\text{height}}, \text{object}_{\text{width}})/4$ yields correct results.

Color Domain Updating: Once the spatial components of the Gaussians have been updated, we update the foreground model according to the color domain. For each foreground and background Gaussian, we check if the data that it is modeling (according to the pixels assigned to this Gaussian) follows a Gaussian distribution. The multidimensional Kolmogorov-Smirnov test ([18]) can be used for this aim. Otherwise, simple tests based on distances can be applied to the pixels assigned to a Gaussian in order to compute the percentage of these pixels that are well described by the Gaussian.

- If the data follows a Gaussian distribution, only one Gaussian is needed to model these pixels. In this situation, we first analyze whether a color updating is needed, comparing the Gaussian distribution in analysis with the Gaussian distribution that better models the data. This comparison can be made via Kullback-Leibler divergence ([19]) or with simple tests that compare, for pixel i, each component c of the mean values (μ_1 and μ_2) of the two distributions in relation with their variances (σ_1^2 and σ_2^2),

$$\|\mu_{1,c} - \mu_{2,c}\|^2 < min(k^2\sigma_{1,c}^2, k^2\sigma_{2,c}^2), \tag{3}$$

  where k is a decision constant (we use $k = 2.5$). Index 1 and index 2 denote the Gaussian distributions that we want to compare. In this case, index 1 denotes the Gaussian distribution of the foreground model and index 2 denotes the Gaussian distribution that better models the data. If the Gaussian in analysis models correctly the data, no updating is necessary. Otherwise, the color domain parameters of the Gaussian are replaced by the data ones.
- If not, it means that more than one Gaussian is needed to model these pixels. Another Gaussian distribution is created, and we use the EM algorithm ([16]) to maximize the likelihood of the data in the spatial and color domains.

In order to increase the robustness of the system, color updating of the foreground and background model is only performed if the Gaussian of the model in analysis is different enough in the color domain from the Gaussians of the other model that correspond to the same spatial positions. Again, we can apply Kullback-Leibler divergence or compare the mean value of the distributions. For instance, we consider that the foreground model can be updated if at least 70% of the pixels that the new Gaussian represents have a background model that does not accomplish (3).

3.3   Classification

Once the foreground and background models have been computed at frame $t-1$, the labeling can be done for the frame t assuming that we have some knowledge of foreground and background prior probabilities, $P(\text{fg})$ and $P(\text{bg})$ respectively, using a Maximum A Posteriori (MAP) decision. The priors can be approximated by using the foreground and background areas, computed as number of pixels, in the previous frame, $t - 1$,

$$P(\text{fg}) = \frac{Area_{\text{fg}}|_{t-1}}{N}; \quad P(\text{bg}) = \frac{Area_{\text{bg}}|_{t-1}}{N};$$

A pixel i may be assigned to the class $l_i = \{\text{fg,bg}\}$ that maximizes $P(l_i|z_i) \propto P(z_i|l_i)P(l_i)$ (since $P(z_i)$ is the same for all classes and thus can be disregarded).

Analogously to [12,13,14] we choose to consider the spatial context also for taking the segmentation decisions, instead of making an individual classification of the pixels. We consider for this aim a MAP-MRF framework in order to take into account neighborhood information. Then, if we denote by l the labeling of all pixels of the image: $l = \{l_1, l_2, ..., l_N\}$, and by Nb_i the four connected neighborhood of pixel i:

$$P(l|z) \propto \prod_{i=1}^{N} P(z_i|l_i)P(l_i)e^{\sum_{i=1}^{N}\sum_{j \in Nb_i} \lambda(l_i l_j + (1-l_i)(1-l_j))} \qquad (4)$$

Taking logarithms in the above expression leads to a standard form of the energy function that is solved for global optimum using a standard graph-cut algorithm ([20]).

4   Results

This section shows some tests to evaluate the quality and robustness of the system proposed [1]. For this purpose, qualitative and quantitative evaluations have been performed. Quantitative results are obtained analyzing the cVSG public Data Base [21]. Qualitative results are obtained analyzing another two different video sequences with different difficulty degree. In Figure 4 the shirt of a running girl has been segmented. These results show how the shirt is correctly detected along the sequence despite the variability of the background regions. Moreover, in this sequence the evolution of the spatial foreground and background models along the sequence can be observed. Each ellipse is the graphical representation of each Gaussian distribution.

Figure 5 shows the results obtained in a F1 sequence that presents special difficulty due to object translation and rotation and the presence of other similar F1 cars within the area of analysis. It can be observed how the proposed system achieves a correct and robust object segmentation over these conditions, and adapts well to all these new regions that appear within the Dynamic Region of Analysis in each frame. Thanks to background model color and spatial updating, new background regions that appear in each frame, are incorporated into the background model before they affect the foreground model.

Table 1 shows the quantitative results using cVSG public database [21]. This database presents several sequences with different difficulty degree, depending on the background characteristics and the foreground to segment. We have used the full length of each sequence to compute the numerical results. The metrics used in the evaluation are: Precision (P), Recall (R) and f_{measure} metrics, formulated as follows:

$$P = \frac{TP}{TP + FP}; \qquad R = \frac{TP}{TP + FN}; \qquad f_{\text{measure}} = \frac{2RP}{R + P}.$$

---

[1] Available in: https://imatge.upc.edu/~jgallego/MUSCLE2011/results/

Table 1. Quantitative Results using cVSG Public Data Base [21]

Sequence	Precision	Recall	$f_{measure}$
Dancing (v.1 boy)	0.942	0.988	0.965
Dancing (v.1 girl)	0.934	0.992	0.962
Dancing (v.2 boy)	0.985	0.990	0.987
Dancing (v.2 girl)	0.984	0.992	0.988
Dangerous race	0.958	0.994	0.975
Exhausted runner	0.986	0.985	0.986
Bad manners	0.978	0.991	0.984
Teddy bear	0.916	0.981	0.948
Hot day	0.980	0.985	0.983
Playing alone	0.997	0.984	0.990

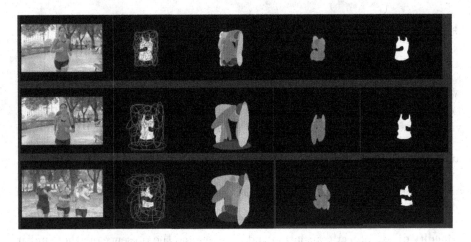

Fig. 4. Results. Girl sequence. From left to right: original image, resultant mask with the Gaussians corresponding to spatial representation of the foreground model (red) and the background model (green), spatial representation of the background model (each Gaussian is colored with the mean color that it is modeling), spatial representation of the foreground model(each Gaussian is colored with the mean color that it is modeling), resultant foreground object mask.

where TP, FP and FN are *TruePositive*, *FalsePositive* and *FalseNegative* pixels detected in the evaluation: frame, sequence or set of sequences.

As it can be observed, the system proposed achieves a high $f_{measure}$ score in the overall data base although moving and dynamic background are present.

Regarding the computational cost, the system allows a speed of 1 frames/ second, for a video sequence of 720x576 pixels with one object in scene, and using an Intel Xeon X5450 3.0GHz processor.

Fig. 5. Results. F1 sequence. From left to right and from top to bottom: original image and the resultant object mask.

5   Conclusions

We have proposed in this paper a novel foreground segmentation system for moving camera scenarios, based on the use of the region-based spatial-color GMM to model the foreground object to segment and moreover, the close background regions that surrounds the object. We have proposed a motion framework for these kind of sequences that has allowed us to associate the camera motion to the background plane and to consider the probabilistic modeling of these close-background regions to achieve the classification of the pixels inside the ROI into the foreground and background classes within a MAP-MRF framework. The results show that the proposed system achieves a correct object segmentation reducing the false positives, and false negatives detections also in those complicated scenes where camera motion, object motion and camera zoom are present, as well as similarity between foreground and background colors.

References

1. Wren, C.R., Azarbayejani, A., Darrell, T., Pentland, A.P.: Pfinder: Real-time tracking of the human body. IEEE Trans. on Pattern Analysis and Machine Intelligence 19(7), 780–785 (2002)
2. Stauffer, C., Grimson, W.E.L.: Adaptive background mixture models for real-time tracking. In: IEEE Proc. Computer Society Conference on Computer Vision and Pattern Recognition, vol. 2, pp. 246–252 (1999)
3. Isard, M., Blake, A.: Condensation conditional density propagation for visual tracking. International Journal of Computer Vision 29(1), 5–28 (1998)
4. Cristani, M., Farenzena, M., Bloisi, D., Murino, V.: Background subtraction for automated multisensor surveillance: a comprehensive review. EURASIP Journal on Advances in Signal Processing 2010, Article ID 343057, 24 (2010)

5. Araki, S., Matsuoka, T., Yokoya, N., Takemura, H.: Real-time tracking of multiple moving object contours in a moving camera image sequence. IEICE Trans. on Information and Systems 83(7), 1583–1591 (2000)
6. Sawhney, H.S., Ayer, S.: Compact representations of videos through dominant and multiple motion estimation. IEEE Trans. on Pattern Analysis and Machine Intelligence 18(8), 814–830 (2002)
7. Jin, Y., Tao, L., Di, H., Rao, N.I., Xu, G.: Background modeling from a free-moving camera by multi-layer homography algorithm. In: Proc. IEEE International Conference on Image Processing, pp. 1572–1575 (2008)
8. Smith, S.M., Brady, J.M.: ASSET-2: Real-time motion segmentation and shape tracking. IEEE Trans. on Pattern Analysis and Machine Intelligence 17(8), 814–820 (2002)
9. Grinias, I., Tziritas, G.: A semi-automatic seeded region growing algorithm for video object localization and tracking. Signal Processing: Image Communication 16(10), 977–986 (2001)
10. Cucchiara, R., Prati, A., Vezzani, R.: Real-time motion segmentation from moving cameras. Real-Time Imaging 3(10), 127–143 (2004)
11. Leichter, I., Lindenbaum, M., Rivlin, E.: Bittracker, a bitmap tracker for visual tracking under very general conditions. IEEE Trans. on Pattern Analysis and Machine Intelligence 30(9), 1572–1588 (2008)
12. Yu, T., Zhang, C., Cohen, M., Rui, Y., Wu, Y.: Monocular video foreground/background segmentation by tracking spatial-color Gaussian mixture models. IEEE Workshop on Motion and Video Computing (2007)
13. Sheikh, Y., Shah, M.: Bayesian modeling of dynamic scenes for object detection. IEEE Trans. on Pattern Analysis and Machine Intelligence 27(11), 1778–1792 (2005)
14. Gallego, J., Pardas, M., Haro, G.: Bayesian foreground segmentation and tracking using pixel-wise background model and region based foreground model. IEEE Int. Conf. on Image Processing, 3205–3208 (2009)
15. Gallego, J., Pardas, M.: Enhanced bayesian foreground segmentation using brightness and color distortion region-based model for shadow removal. In: IEEE Int. Conf. on Image Processing, pp. 3449–3452 (2010)
16. Dempster, A.P., Laird, N.M., Rubin, D.B.: Maximum likelihood from incomplete data via the EM algorithm. Journal of the Royal Statistical Society. Series B (Methodological) 39(1), 1–38 (1977)
17. Khan, S., Shah, M.: Tracking people in presence of occlusion. In: Asian Conf. on Computer Vision, vol. 5 (2000)
18. Fasano, G., Franceschini, A.: A multidimensional version of the Kolmogorov-Smirnov test. Royal Astronomical Society 225, 155–170 (2000); Monthly Notices (ISSN 0035-8711)
19. Kullback, S.: The kullback-leibler distance. The American Statistician 41, 340–341 (1987)
20. Boykov, Y., Veksler, O., Zabih, R.: Fast approximate energy minimisation via graph cuts. IEEE Trans. on Pattern Analysis and Machine Intelligence 29, 1222–1239 (2001)
21. Tiburzi, F., Escudero, M., Bescós, J., Martínez, J.M.: A Ground-truth for Motion-based Video-object Segmentation. In: IEEE Int. Conf. on Image Processing Workshop on Multimedia Information Retrieval: New Trends and Challenges, pp. 17–20 (2008)

Real Time Image Analysis for Infomobility

Massimo Magrini, Davide Moroni, Gabriele Pieri, and Ovidio Salvetti

Institute of Information Science and Technologies (ISTI),
National Research Council of Italy (CNR), Pisa, Italy
name.surname@isti.cnr.it

Abstract. In our society, the increasing number of information sources is still to be fully exploited for a global improvement in urban living. Among these, a big role is played by images and multimedia data (i.e. coming from CCTV and surveillance videos, traffic cameras, etc.). This along with the wide availability of embedded sensor platforms and low-cost cameras makes it now possible the conception of pervasive intelligent systems based on *vision*. Such systems may be understood as distributed and collaborative sensor networks, able to produce, aggregate and process images in order to understand the observed scene and communicate the relevant information found about it. In this paper, we investigate the characteristics of image processing algorithms coupled to visual sensor networks. In particular the aim is to define strategies to accomplish the tasks of image processing and analysis over these systems which have rather strong constraints in computational power and data transmission. Thus, such embedded platform cannot use advanced computer vision and pattern recognition methods, which are power consuming, on the other hand, the platform may be able to exploit a multi-node strategy that allows to perform a hierarchical processing, in order to decompose a complex task into simpler problems. In order to apply and test the described methods, a solution to a visual sensor network for infomobility is proposed. The experimental setting considered is two-fold: acquisition and integration of different views of parking lots, and acquisition and processing of traffic-flow images, in order to provide a complete description of a parking scenario and its surrounding area.

1 Introduction

The application of computer vision methods to the large and heterogeneous amount of information available nowadays is a tricky and complex goal. The first thing to take into account is the platform for which these methods need to be implemented. When a large and powerful platform is available (e.g. desktop PC-like, clouds,...) many different and more powerful methods and algorithms can be implemented. On the other hand, aiming to achieve a low-cost, low-consumption and pervasive implementation, platforms like embedded systems need to be considered.

Recently, embedded systems have become widely available along with low-cost camera sensors, thus allowing to design sensor-based intelligent systems centered on image data [1]. These vision systems can be connected wireless in networks, forming the so called *visual Wireless Sensor Networks* (WSNs). Such systems, formed by a large

E. Salerno, A.E. Çetin, and O. Salvetti (Eds.): MUSCLE 2011, LNCS 7252, pp. 207–218, 2012.

number of nodes each carrying a low power camera, may be important in supporting novel vision-based applications, considering all the information that can be retrieved out of images. Visual WSNs are often used for the real-time monitoring of large and full of obstacles areas, like shopping malls, airports and stadiums. Image mining of the acquired scenes through the network can be used for the detection of statistically significant events [2]. Moreover, visual WSNs can successfully be applied to environmental monitoring, to the remote control of elderly patients in e-Health [3], and also in ambient intelligence applications (e.g. human gesture analysis) [4].

The abundance of data available and all the information that could be extracted through image analysis and understanding methods allows for the use of WSNs in a very prominent way. This means that a visual WSN is to be intended as a network of sensors spread over a specific area, which communicate and collaborate to mine the scene under investigation for detecting, identifying and describing specific characteristics. Each sensor is called *node* and owns, at the same time, acquisition and analysis capabilities. Moreover each node is designed in order to be able to transmit and collaborate with the other nodes in order to reach an *aggregated result* from all the different views and analysis of the scene.

A successful design and development of such a system cannot be achieved without suitable solutions to the involved computer vision problems. Although the computer vision problems may be still decomposed into basic computational tasks (such as feature extraction, object detection and object recognition), in the context of visual WSNs, it is not directly possible to use all the state-of-the-art methods, which often are tailored to other, more powerful, architectures [5].

One of the main features of WSNs is the large number of nodes that can be involved, spread over a possible large area. Thus in order to make such a solution achievable, the cost per node needs to be modest. As a consequence, each node has to be implemented with limited computational power, and also with low resolution cameras, thus setting constraints also on the implementation of the computer vision methods.

Hence, the opportunity offered by WSNs to perform a pervasive computing in an efficient way is balanced by the challenging goal of developing suitable (in terms of technological constraints) computer vision methods able to exploit the organized network topology and the hierarchy of nodes for an effective mining and investigation of a scene.

This paper describes a possible solution to a multi-node processing task based on the decomposition of more complex vision tasks into a hierarchy of computationally simpler problems, to be solved over the nodes of the network. This solution is suitable not only for the computational constraint that it respects, but also because such a hierarchical approach to a decision making task brings to a more robust and fault-tolerant solution. The solution has been applied and tested to a real infomobility case study.

We considered two different set-ups of the visual WSN; the first one consists of several different views of a parking lot, integrated through the communication among the different nodes; the second one has been designed to acquire and integrate real time images for traffic flow control, thus having harder timing constraint. The methods implemented for this solution permitted to aggregate such information, in order to provide a description of the case study in terms of localization and count of available spaces, and to analyze and count the number of vehicles in various roads used to access the parking surrounding area. A prototype of such visual WSN was effectively installed on a test-bed and data were collected for performance and validation purposes, thus showing that the proposed paradigms are not only valid from a theoretical point-of-view but also promising for deployment in real world applications. With respect to [6] and [7] where the problem was originally defined and addressed, in this paper we refine the employed methods and provide a richer set of experimental results.

2 Background

Following the trends in low-power processing, wireless networking and distributed sensing, visual WSNs are experiencing a period of great interest, as shown by the recent scientific production (see e.g [8]). A visual WSN consists of tiny visual sensor nodes called camera nodes, which integrate the image sensor, the embedded processor and a wireless RF transceiver [1]. The large number of camera nodes forms a distributed system where the camera nodes are able to process image data locally (*in-node processing*) and to extract relevant information, to collaborate with other cameras – even autonomously – on the application specific task, and to provide the user with information-rich descriptions of the captured scene.

In this case, instead of a centralized approach, i.e. transferring the images to a central unit and performing the whole processing there, a distributed system – both from a physical and semantic point of view – is well motivated and can be efficient under many different aspects [9]:

- *Speed:* distributed processing is inherently parallel among nodes; in addition, the specialization of modules permits to reduce the computational burden in the high level decisional nodes.
- *Bandwidth:* the processing internal to each node permits to reduce the quantity of transmitted data; the data transmitted can be just parameters about the observed scene and not the redundant image data stream.
- *Redundancy:* a distributed system may be re-configured in case of failure of some of it components, still keeping the overall functionalities.
- *Autonomy:* asynchronous processing; each node independently processes the images and reacts to the detected changes in the scene.

Considering these aspects, the artificial intelligence and computer vision algorithms could be arranged to perform part of the processing along the nodes of the WSN. This *decentralization* of the intelligence allows the system to have autonomy and adaptation to *internal conditions* (e.g. hardware and software failure) as well as to *external conditions* (e.g. changes in weather and lighting conditions).

Each of the involved computer vision issues has been already widely studied and different approaches to their solution have been faced in the literature (see e.g. [10] for a survey of change detection algorithms); unfortunately, most of the techniques have heavy computational or memory requirements, which are not compatible to address the same problems when operating within a visual WSNs. However, in literature some efforts have been made to use image processing methods with a visual WSN. In [11] the visual WSN used is able to support the query of a set of images, in order to search for a specific object in the scene, through a representation of the object following the Scale Invariant Feature Transform (SIFT) descriptors [12]. SIFT descriptors are known to support robust identification of objects even among cluttered background and under partial occlusion situations, since the descriptors are invariant to scale, orientation, affine distortion and partially invariant to illumination changes. In particular, these descriptors allow to detect the object of interest in the scene, independently from the scale it is imaged.

Reviewing a specific vision system, the CMUcam3 [13] is also supplied with both basic image processing filters (such as convolutions), and methods for real-time tracking of blobs on the base of either color homogeneity or frame differencing. Furthermore, a customizable face detector is also included, based on a simplified implementation of the Viola-Jones detector [14], enhanced with some heuristics (e.g. avoid searching in low variance region of the image) to further reduce the computational burden.

As mentioned above, the camera sensors mounted on WSNs usually have limited resolution and field of view, thus in order to be able to elaborate and mine an entire scene, possibly of a large area, and to deal with occlusions, a multi-view approach needs to be implemented [15]. This approach is taken into account considering that it is not possible, for efficiency aspects, to exploit features like image sharing through the network (e.g. for computing 3D properties like depth maps). Despite this, some constraints can be set, which do not reduce the applicability of the solution. For example geometric objects which are static in the scene could be codified into the nodes at set-up time, along with the use of visual references existing or created into the scene, for a calibration of the views by groups of nodes. In this sense, a correspondence among all the nodes in a topological and geometrical map can be obtained. In order to improve this, one of the nodes in the hierarchy of the network is to be referred to as the *coordinator*, i.e. a node that is aware of needed translation between single nodes coordinates and real world ones. This solution is also important for the objective of guaranteeing more robust results and an information-rich description of the scene to be mined.

3 Methods

In a visual WSN, the goal of the system is to report any relevant change in the scene, where some real-time constraints are to be satisfied. To this end, one should take into account both local computation and network communication issues to bound the maximum latency in the reaction.

Considering these constraints, but in the need for an efficient mining of the visual scene under examination, we focused on two different approaches: the first one based on simpler detection methods, but more efficient from a computational and storage points of view, the second giving better performances for more specific object identification tasks while being computationally more intensive.

Moreover, considering a more global analysis and mining of the scene viewed from several different nodes (i.e. multi-view), a final processing level for the aggregation of different single in-node data and the final result of the scene mining needs to be implemented.

In the following, we analyze the above mentioned specific methods for the detection of changes in a scene, in particular regarding the two categories: simple but quick and efficient change detection algorithms in Section 3.1, and more complex and demanding algorithms for an object detection and identification in Section 3.2. Finally, in Section 3.3 we present methods for the aggregation of hierarchical identifications, made at different levels of complexity and from multiple acquiring points of view, in order to be able to correctly process the set of images to detect events.

3.1 Event and Change Detection

The task of identifying changes, and in particular regions in the image under analysis, taken at different times is a widely diffused topic encompassing very different disciplines [10]. In this context, the goal is the identification of one or more regions relative to a change in the scene and occurring within a sensible and a priori known area.

In order to perform fast, efficient and reliable methods for change detection, the generally adopted low-level methods are based on both frame differencing and statistical methods. This can be either quick and immediate answer to simple problems, or a preliminary synthesis for a deeper and more effective higher level analysis. The general model used is mainly based on background subtraction, where a single frame, acquired in a controlled situation, is stored (*BG-image*) and thereafter used as the zero-level, for different types of comparison in the actual processing.

The first methods are very low-level ones and are thought so that it is possible and feasible to implement them also at firmware level, so to make them real-time operative in an automatic way; for example, the following methods were experimented:

- Thresholding toward the BG-image
- Contrast modification with respect to the BG-image
- Basic edge detection on the image resulting from the subtraction with the BG-image

Another class of methods is based on histogram analysis. Template histograms are stored of the regions of interest in the BG-image. Then a standard distance, such as the Kullback-Leibler divergence, is computed between the region of BG-image, and the region of the actual image. If such a distance exceeds a fixed threshold, then a change event is detected.

3.2 Object Detection

As is well-known, the problem of detecting classes of objects in cluttered images is challenging. Supervised learning strategies have demonstrated to provide a solution in a wide variety of real cases, but there is still a strong research in this field. In the context of visual WSN, the preliminary learning phase may be accomplished off-site, while only the already trained detectors need to be ported to the network nodes.

Among machine learning methods, a common and efficient one is based on the sliding-windows approach. A priori knowledge about the scene and information already gathered by other nodes in the network may be employed to reduce the search space either by a) disregarding some region in the image and b) looking for rectangular regions

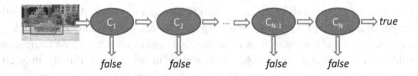

Fig. 1. Cascade of classifiers for object detection in a case study of car detection

within a certain scale range. For what regards the binary classifiers, among various possibilities, the Viola-Jones method is particularly appealing. Indeed, such a classifier is based on the use of the so-called Haar-like features, a class of features with limited flexibility but known to support effective learning. Haar-like features are computable in constant time, once the integral image has been computed (see [14] for details). Thus, having enough memory to store the integral image, the feature extraction process needs only limited computational power. Classification is then performed by applying a cascade of classifiers, as shown in Figure 1. A candidate subwindow which fails to meet the acceptance criterion in some stage of the cascade is immediately rejected and no further processed. In this way, only *detection* should go through the entire cascade.

The use of a cascade of classifiers permits also to adapt the response of the detector to the particular use of its output in the network, also in a dynamical fashion, in order to properly react to changes in the internal and external conditions. First of all, the trade-off between reliability and computational time may be controlled by adaptive real-time requirements of the overall network. Indeed, the detector may be interrupted at an earlier stage in the cascade, thus producing a quick, even though less reliable, output that may be anyhow sufficient for solving the current decision making problem. In the same way, by controlling the threshold in the last stage of the cascade, the visual WSN may dynamically select the optimal trade-off between false alarm rate and detection rate needed in a particular context.

3.3   Aggregation of Partial Results

Considering a visual WSN with multiple views of the same scene for image mining purposes, the previously described methods can be managed in a hierarchical fashion, with faster algorithms yielding a primary response to a simple change detection task, and sending their results up through the hierarchy to more performing algorithms, in case of a positive detection. These higher level algorithms, due to their greater computational requirements, are implemented in order to be called only when positive feedback is given by the lower level ones. At the highest level of the hierarchy, a final *decision-maker* algorithm operates without strict computational and storage constraints for aggregating the results of lower level processing and combining the multiple views (see Fig. 2).

A common hypothesis for the multiple view aggregation process works on the basis of a priori information, relative to the approximate possible positions, or Regions of Interest, where the objects to be analyzed can be detected. For example the decision maker algorithm can analyze the specific outcomes from different lower-level

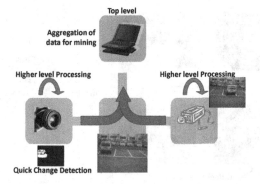

Fig. 2. Architecture of the hierarchical levels of detection, in a car detection case study

processing in each single node, and then be able to give a final output using its knowledge and working on the basis of a weighted computation of the single in-node results.

4   Case Study: A Pervasive Infrastructure for Urban Mobility

After having presented methods for performing image analysis over WSN, in this section we describe a real case study concerning urban mobility. To this end, we describe the embedded vision sensor prototypes to be used in the project and, finally, we report preliminary results in a particular scenario.

The Tuscany regional project IPERMOB – "A Pervasive and Heterogeneous Infrastructure to control Urban Mobility in Real-Time" [16] is exploiting an approach based on visual WSNs for building an intelligent system capable of analyzing infomobility data and of providing effective decision support to final users, e.g. citizens, parking managers and municipalities.

In IPERMOB, visual WSNs are used to monitor parking lots and roads, in order to mine the observed scenes and predicate something about parking lot occupancy and traffic flow. To this end, a number of embedded sensor prototypes is being considered. In particular, two different architectures are being explored:

- a dual-microcontroller architecture, in which image acquisition and processing functions are separated from the network communication operations;
- a FPGA based board with dedicated hardware components for image processing and a soft microprocessor for image acquisition control and wireless communication.

Different scenarios for the evaluation of both parking slots and traffic flow have been set-up. For the parking lot scenario, a case study at the parking lot of Pisa International Airport was used. The set-up consists in a set of WSN nodes equipped with cameras with partially overlapping field of views. The goal was to observe and estimate the availability and location of parking spaces. A basic assumption was made on the geometry of the parking: each camera knows the positions of the parking slots under its monitoring. In addition, we assume that a coordinator node knows the full geometry of

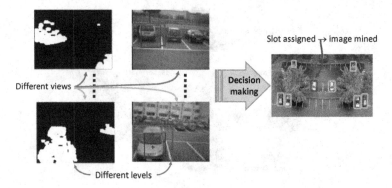

Fig. 3. The aggregation and decision-making process for WSN image mining in IPERMOB Project

Fig. 4. Car detection and analysis of parking lot occupancy status

the parking lot as well as the calibration parameters of the involved cameras, in order to properly aggregate their outcomes. In figure 3, the workflow in the real case study is reported.

For the traffic flow, the set-up consists in a smaller set of WSN nodes, still equipped with cameras which is in charge of observing and estimating dynamic real-time traffic related information, in particular regarding traffic flow and the number and direction of the vehicle, as well as giving a rough estimate about the average speed of the cars passing by.

5   Results

First of all, the results regarding the solution proposed for the parking lot analysis are reported. In a typical scenario, the case study under investigation, some of the camera nodes in the first level of the hierarchy detect a change in the scene (by the methods reported in Section 3.1) and trigger object detection methods in the remaining nodes in charge of monitoring that area. The aggregation and decision making process are finally performed by the coordinator node.

For what regards object detection, several cascades of classifiers have been trained, each specialized in detecting particular views of cars (e.g. front, rear and side views). A large set of labeled acquisitions have been made of the real case study and used for training purposes. We notice that the first stages in the cascade have low computational

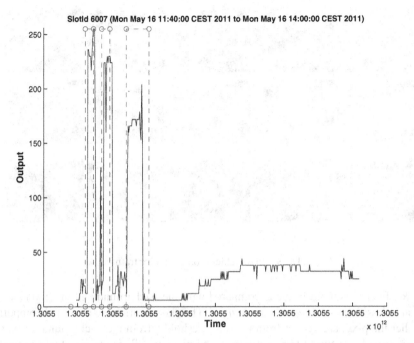

Fig. 5. Data collected for a parking slot on the case study site. Sensor confidence values are shown in blue, while the ground-truth recorded is shown in red with circles representing change-event.

complexities but reasonable performances (i.e. almost 100% detection rate but even 50% false alarm rate). Composing the stages entails a high performance of the entire cascade, since both overall detection rate and overall false alarm rate are just the products of the detection rate and false alarm rate of the single stages. For example, using $N = 10$ stages in the cascade, each with detection rate $v = 99.5\%$ and false alarm rate $\mu = 50\%$, one gets an overall detection rate $v_{global} = v^N \approx 95\%$ and false alarm rate $\mu_{global} = \mu^N \approx 0.1\%$.

Figures 4 and 6 show examples of processed image sequences, respectively for parking availability and traffic flow. Figure 5 shows the differences between the sensor values and the ground-truth recorded for a parking slot acquisition, that shows the good separation obtained between different events.

Regarding the traffic flow monitoring, two versions of the algorithm were implemented. In the first, the solutions used three different frames (using frame differencing) to obtain a binary representation of the moving objects in the reference frame. Analyzing the connected components, blobs are detected, and then it is verified whether these can be referred to objects moving through (i.e. traffic flow) a predefined Region of Interest (RoI). The final version of the algorithm was designed to eliminate the analysis of connected components, considering that from the first experimental tests this analysis could be too time consuming for the available hardware. In particular, the algorithm uses a RoI R_k for each flow measure Φ_k ($k = 1, 2, \ldots$) to be computed and the number of vehicle flowing through is analyzed. A background image B_k is stored for each

Fig. 6. Detection of vehicles on a road for traffic flow

RoI R_k. Every acquired frame is compared with the background and the pixels in the RoI are classified as *changed* or *unchanged*. The classification is performed computing whether the pixel are distant from a fixed threshold γ from the background. In particular, if I_t is the acquired frame at time t, the pixel $p \in R_k$ is classified as changed if $|I_t(p) - B_k(p)| \geq \gamma$. Furthermore $a_k(t)$ is computed as:

$$a_k(t) = \frac{\#(\text{changed pixels in } R_k)}{\#(\text{pixels in } R_k)}$$

corresponding to the fraction of pixel changed with respect to the total number of pixel in the RoI. Then $a_k(t)$ is compared to a threshold τ in order to evaluate if a vehicle was effectively passing by R_k.

Once a traffic flow has been detected on a RoI R_k, the detection of additional vehicles in the same RoI is inhibited for a fixed number of frames. In order to make the algorithm more robust with respect to variation in the scene (e.g. temporary occlusions, pedestrian passage, ...), the background was chosen to be adaptive. In detail, the background is updated every frame *fusing* the actual with the previous frame:

$$B_k(p) \leftarrow \alpha_t I_t(p) + (1 - \alpha_t)B_k(p)$$

where $0 \leq \alpha_t \leq 1$. The parameter α_t represents the background update speed at time t. Heuristically a higher value η_1 of the update speed was chosen when $a_k(t) < \tau$ (i.e. no vehicles on the RoI R_k) while a lower value η_2 is chosen when $a_k(t) \geq \tau$, meaning that a vehicle could be actually on the region R_k.

$$\alpha_t = \begin{cases} \eta_1 \text{ if } a_k < \tau \\ \eta_2 \text{ if } a_k \geq \tau \end{cases}$$

with $\eta_1 > \eta_2$.

These values were empirically set to: $\eta_1 = 0.2$ and $\eta_2 = 0.05$.

In table 1, the results of the traffic flow case study are reported, showing the improvement in performance between the preliminary version and the final implemented solution.

Table 1. Traffic flow analysis algorithm performance comparison between preliminary and final version

	Sequence	Hit	Miss	False positive	Total real events	Sensitivity rate	False positive
	Sequence 1	204	24	9	228	89%	4%
V1	Sequence 2	234	2	10	236	99.2%	4.2%
	TOTAL	**438**	**26**	**19**	**464**	**94.3%**	**4.1%**
	Sequence 1	226	2	3	228	99.1%	1.3%
V2	Sequence 2	234	2	2	236	99.2%	0.8%
	TOTAL	**460**	**4**	**5**	**464**	**99.1%**	**1.1%**

6   Conclusion and Further Work

The image mining problem has been tackled in this paper, through the use of visual sensor networks. Structural limits of embedded platforms have been deeply taken into account, balanced with the benefit of a solution organized implementing pervasive computing and a flexible network topology.

The proposed solution is based on multi-node processing, in order to decompose complex computer vision tasks into a hierarchy of simpler problems, both from a computational and processing points of view, to be solved over the nodes of the network. Besides computational advantages, such a hierarchical decision making approach appears to be more robust and fault-tolerant.

Robustness is given by both multi-node and multi-view design of the sensor network, with same areas covered by more than one sensor, thus guaranteeing that the outcome is balanced and weighted between different points of view, and that if a single node fails due to a hardware problem, its RoI will still be mostly covered.

These ideas and solution were finally applied on a case-study in infomobility, developed within the frame of the IPERMOB project. After the preliminary testing, the solution implemented seems promising and goes into the direction of proving that such a solution is mature for technological transfer. Finally, as shown by the results presented in the above tables, the successive and refined versions of the algorithms give increased and very encouraging performances.

Acknowledgments. This work has been partially supported by POR CReO Tuscany Project "IPERMOB" – A Pervasive and Heterogeneous Infrastructure to control Urban Mobility in Real-Time (Regional Decree N. 360, 01/02/2010). We would like to thank our colleagues from ReTis Lab (CEIIC) at Scuola Superiore Sant'Anna in Pisa for their contribution in a large part of the work, and also Dr. Mario De Pascale (IT Manager at Società Aeroporto Toscano S.p.A.) for his kind support and for providing access to the testing area.

References

1. Soro, S., Heinzelman, W.: A survey of visual sensor networks. Advances in Multimedia, Article ID 640386, 21 (2009)
2. Adam, A., Rivlin, E., Shimshoni, I., Reinitz, D.: Robust real-time unusual event detection using multiple fixed-location monitors. IEEE Trans. PAMI 30, 555–560 (2008)
3. Colantonio, S., Conforti, D., Martinelli, M., Moroni, D., Perticone, F., Salvetti, O., Sciacqua, A.: An intelligent and integrated platform for supporting the management of chronic heart failure patients. Computers in Cardiology, 897–900 (2008)
4. Salvetti, O., Cetin, E.A., Pauwels, E.: Special issue on human-activity analysis in multimedia data. Eurasip Journal on Advances in Signal Processing, article n. 293453 (2008)
5. Pagano, P., Piga, F., Lipari, G., Liang, Y.: Visual tracking using sensor networks. In: Proc. 2nd Int. Conf. Simulation Tools and Techniques, ICST, pp. 1–10 (2009)
6. Magrini, M., Moroni, D., Nastasi, C., Pagano, P., Petracca, M., Pieri, G., Salvadori, C., Salvetti, O.: Image mining for infomobility. In: 3rd International Workshop on Image Mining Theory and Applications, pp. 35–44. INSTICC Press, Angers (2010)
7. Magrini, M., Moroni, D., Nastasi, C., Pagano, P., Petracca, M., Pieri, G., Salvadori, C., Salvetti, O.: Visual sensor networks for infomobility. In: Pattern Recognition and Image Analysis, pp. 20–29 (2011)
8. Kundur, D., Lin, C.Y., Lu, C.S.: Visual sensor networks. EURASIP Journal on Advances in Signal Processing Signal Processing 2007, Article ID 21515, 3 (2007)
9. Remagnino, P., Shihab, A.I., Jones, G.A.: Distributed intelligence for multi-camera visual surveillance. Pattern Recognition 37, 675–689 (2004)
10. Radke, R.J., Andra, S., Al-Kofahi, O., Roysam, B.: Image change detection algorithms: a systematic survey. IEEE Transactions on Image Processing 14, 294–307 (2005)
11. Yan, T., Ganesan, D., Manmatha, R.: Distributed image search in camera sensor networks. In: Abdelzaher, T.F., Martonosi, M., Wolisz, A. (eds.) SenSys, pp. 155–168. ACM (2008)
12. Lowe, D.G.: Distinctive image features from scale-invariant keypoints. International Journal of Computer Vision 60, 91–110 (2004)
13. Rowe, A., et al.: CMUcam3: An open programmable embedded vision sensor. Technical Report CMU-RI-TR-07-13, Robotics Institute, Pittsburgh, PA (2007)
14. Viola, P.A., Jones, M.J.: Robust real-time face detection. International Journal of Computer Vision 57, 137–154 (2004)
15. Pagano, P., Piga, F., Liang, Y.: Real-time multi-view vision systems using WSNs. In: Proc. ACM Symp. Applied Comp., pp. 2191–2196. ACM (2009)
16. IPERMOB: A Pervasive and Heterogeneous Infrastructure to control Urban Mobility in Real-Time (2010), http://www.ipermob.org/ (Last retrieved February 8, 2012)

Tracking the Saliency Features in Images Based on Human Observation Statistics

Szilard Szalai[1,2], Tamás Szirányi[1,2], and Zoltan Vidnyanszky[1]

[1] Pázmány Péter Catholic University, Faculty of Information Technology,
Budapest, Hungary
{szasz,vidnyanszky}@digitus.itk.ppke.hu
[2] Distributed Events Analysis Research Group,
Computer and Automation Research Institute,
Hungarian Academy of Sciences (MTA SZTAKI),
Budapest, Hungary
sziranyi.tamas@sztaki.mta.hu

Abstract. We address the statistical inference of saliency features in the images based on human eye-tracking measurements. Training videos were recorded by a head-mounted wearable eye-tracker device, where the position of the eye fixation relative to the recorded image was annotated. From the same video records, artificial saliency points (SIFT) were measured by computer vision algorithms which were clustered to describe the images with a manageable amount of descriptors. The measured human eye-tracking (fixation pattern) and the estimated saliency points are fused in a statistical model, where the eye-tracking supports us with transition probabilities among the possible image feature points. This HVS-based statistical model results in the estimation of possible tracking paths and region of interest areas of the human vision. The proposed method may help in image saliency analysis, better compression of region of interest areas and in the development of more efficient human-computer-interaction devices.

Keywords: eye-tracking, fixation pattern, saliency, descriptors, human observation.

1 Introduction

In the development of automatic Human-Computer Interaction (*HCI*) systems, finding the focus of visual attention of human observers without any restraint has an important role. These automatic interfaces might be important in traffic control, tourist information panels, adaptive advertisement tables, etc.

One solution is the tracking of wandering people by a surveillance camera, determining where a person is looking when his movement is unconstrained. In [16] it has been solved by a Reversible-Jump Markov Chain Monte Carlo sampling scheme with a Gaussian Mixture Model and Hidden Markov Model using head pose and location information. However, this complex solution could

E. Salerno, A.E. Çetin, and O. Salvetti (Eds.): MUSCLE 2011, LNCS 7252, pp. 219–233, 2012.

give limited information about the gaze direction: it determined if a person was looking at the advertisement or not.

In [6] the Esaliency Algorithm is introduced, consisting a pre-attentive segmentation and an attentive graphical model approximation phase. The probability that an image part is of interest (saliency) was defined using some underlying assumptions: the number of target candidates was small, there was a correlation between visual similarity and target-nontarget labels and natural scenes are composed of clustered structural units. This algorithm started with a pre-selection of possible candidates using some segmentation processes and assigning initial probability for them, and then a similarity based Bayesian network was built, finding the most likely joint assignment. In the processing period, the estimated salient objects were determined by using the spanning tree of graphs. For the verification, real salient objects were annotated by humans as ground truth in the training phase.

Eye tracking systems may support us with a partly unsupervised training method of human visual attentional selection. Here the human visual activity can be annotated with fixation points of eye movement while the scene is recorded in a video in parallel. This annotated record can be used to describe the image content as the human apprehends it for two main research directions: to characterize the image content or to characterize the viewer.

The first task concerns machine vision, where the visual preferences of human attention are built in the image features in some training procedure. The second task refers to the computational modeling of the perceptual and neural mechanisms of the human vision. This paper addresses the first task: our goal is to discover the artificial saliency features of images with the statistical properties trained from human eye movement recording experiments. The path of the human gaze, the time of fixation on a selected region and the frequency of switches between the different parts of the images can be typical for an image as well.

There are several technical solutions and research approaches for estimating the path of the human gaze. The advertisement industry is interested in models specialized for different specific tasks, for instance inferring gaze path on web pages [17]. In recent studies [18], [8] the goal is to find numerical correlation between different factors of vision. According to human eye-tracker studies, the work in [18] is based on the discovered fact that gaze fixations are mostly in the center of the images. Moreover, top-down features can be categorized into a strong and a weak group, and these types can influence the viewing behavior in a different way; according to [8], the two groups have different roles in vision. Other research teams analyze the images specifically according to their structures [6], [12]. Noise effects in the visual perceptions are investigated in [7], finding noise-related neural processes when dissociating the effect of noise on sensory processing from that of the processes associated with overall task difficulties by recording EEG and measuring fMRI responses separately in the same task conditions.

In our hypothesis, we assume that the transitions between two consecutive fixation points (eye movements, saccades) are also characteristic of the attention selection processes or the image content; the probability that the eye moves to a given point from another may be characteristic of those points and the features behind them. In our proposed model, any high number of possible salient regions is possible in the scene, contrary to the model assumptions in Esaliency [6], while the task is not to find one (group) of the object(s), but to find the path as the human gaze may connect them. We also have the goal to find the salient regions, but this is based on the dense sections of the probable tracking paths and not on the saliency of the object itself. In our proposed method, we attempt to find the regions of the images that will be selected by visual attention and thus will be fixated by the human observers.

2   The Background of the Proposed Model

In this model, we wanted to combine the top-down and bottom-up methods. The top-down method is integrated by the training videos where a kind of behavior fingerprint about the subject's vision is recorded and used. The bottom-up approach is applied by processing the recorded images by computer vision algorithms. We consider two types of saliency features: in human vision experiments, saliency points are the most attractive regions of the fixation patterns, while the artificial saliency points are the most important characterizing points as image features. The paper describes the fusion of the human and the artificial saliences in one, trained model.

Recent studies [6], [12] mention object or pattern based methods for the detection of saliency. Contrary to these articles, our proposed algorithm uses the connections' graph of low-level salient points; the artificial points are found in a relative fast and simple way by applying the Scale- Invariant Feature Transform (SIFT) developed by David G. Lowe [13]. The usage of a high number of saliency points is more advantageous than using saliency objects of low number, since the task is the annotation of eye tracking.

Our research presents a new approach by applying some basic psychophysical assumptions about the process of human visual exploration and orientation behavior. Personal position information of the human training phase is taken into account: the position and the movement of the head and the distribution of the fixation points on the frames captured from the training videos upgrades the model.

We present a saliency track building algorithm trained by human data, the human training measurements are described in Section 3. The measured human behavior, characterized by statistical biological data and saliency features, is fused to build the human eye movement into a model. Section 4 presents the details of the construction of the model: the processing of the human videos, constructing graphs and giving estimations for the most likely points that could be visited by the human path. After presenting the evaluation psychophysical test experiments, the subjective and the objective results are evaluated in Section 5 by calculating the distance between the computational and the human saliency maps.

Fig. 1. A wearable head-mounted eye-tracker device manufactured by Arrington Research [1]. This frame consists of two cameras and an infra LED. The first camera records the scene in front of the subject, while the other analyzes the movement of the eye. The data of the two cameras are integrated and as a result a video is recorded which contains the actual point of the gaze on each time slot. These records have a fundamental role in our model as the training human data for our algorithm by giving information about the subject's vision. This image is derived from [2].

3   Human Training Experiments

To integrate top-down factors into the model, training experiments were carried out in which the observers' visual attention is considered as top-down subjective visual fingerprints. Although there are many different attractive regions in the scene (in visual sense), visual attention not necessarily directs our gaze into all that regions due to some task-specific reasons. However, we assumed that after recording several videos, top-down-based psychophysical saliency regions can be collected and used for the training.

For the human measurements of fixation patterns, a special wearable, head-mounted eye-tracker device was used (Fig. 1). The GigE-60 [1] [2] is able to monitor and record videos in which the movement of the users' eye is recorded using two cameras. The first camera is watching the right eye flashed by an infra LED revealing the detailed texture of the iris. The other camera, the so-called scene camera is located at the center of the frame. It watches and records the scene located at the front of the subject. We recorded several videos under different conditions, mostly in streets, but some videos were recorded inside buildings as well.

The most important fact was to calibrate the eye-tracker precisely, because this problem depends on the distance to a large extent, since it is unrealistic to use a fixed distance between the moving user and the landmarks. According to experiences, it was better to use larger distance during the calibration than it was typical in the training scene. By the precise calibration, the inaccuracy could be reduced below an appropriate level in the cases when the subjects looked at objects which are either too far or too close. During these human measurements, we had to take into consideration that this device can be strange for its user in the beginning of the tests. In order to improve the quality and the reliability of the results, we recorded some test videos so that the subject gets familiar with the eye tracker.

The reliability of our algorithm depends on the precise arrangement of the human measurements. Video streams were recorded with 30 FPS sampling rate in small (320x240) resolution which was appropriate for our purposes.

Fig. 2. SIFT points clustered by k-means algorithm with supervised initialization. The two images are examples for clustering the images by a partly supervised k-means algorithm, where instead of using randomly generated centers for the clusters, we manually assigned these centers taken from the SIFT points. The different frames captured from the same video show that the same parts were generally clustered into the same clusters, for instance the edges of the steps or the objects on the wall.

4 Eye Tracking Methodology

4.1 Saliency Points Detection and Clustering

The salient features of an image are detected and described by applying SIFT. The generated SIFT points will be allocated to the human interest regions in the fusion phase.

We analyzed the images by our algorithm developed in C++ using the OpenCV library, which was included for the basic task needed during the development. As the resolution of the frame images was small (320 x 240 pixels), we got an average of 300 points on each frame. We followed the clustering of the SIFT points similarly to that in Bag-of-Features methods [9] and [15], by using the k-means. However, to avoid the problem of random initialization in the sparse data representation in the 128 dimensional space, a partly supervised initialization was done by manual center-assignment randomly chosen from the coordinates of the detected SIFT points. After testing the results of the clustering procedure, we found that the modification was favorable since the same points in different image frames taken in the same place were clustered into the same cluster (Fig. 2.). As this figure shows, the clustered SIFT points are able to characterize the images. The optimal value of the clusters was estimated based on the human teaching tests according to our experiences. We discovered that this value depends on the circumstances of the recording and on the subjective features including socialization, interest and feelings. For a few minute-length video, it was enough to define 15-20 clusters; otherwise the values of the transition probability tables were balanced.

Table 1. An example for the measured transition probability tables on the machine defined saliency points. The first row and the first column show the cluster IDs. The table contains the probability of switching from a cluster into another in the training videos. For instace, the probability of swtiching from cluster 13 to cluster 8 is 0.17. This data can be considered as a visual fingerprint about different people. After clustering the salient SIFT points by k-means, the transition probability table is generated for the points of different clusters as the eye moves from one point to another.

	1	2	3	4	5	6	7	8	9	10	11	12	13	14	15
1	0.12	0.09	0.05	0.07	0.09	0.03	0.03	0.09	0.12	0.05	0.03	0.10	0.00	0.05	0.07
2	0.06	0.11	0.05	0.06	0.04	0.03	0.01	0.13	0.09	0.08	0.14	0.06	0.03	0.06	0.04
3	0.06	0.05	0.19	0.05	0.09	0.04	0.03	0.11	0.05	0.08	0.08	0.06	0.04	0.03	0.02
4	0.03	0.11	0.07	0.11	0.06	0.03	0.03	0.06	0.13	0.07	0.07	0.11	0.03	0.04	0.04
5	0.07	0.06	0.11	0.04	0.15	0.01	0.02	0.11	0.12	0.08	0.06	0.07	0.02	0.05	0.01
6	0.07	0.00	0.04	0.18	0.04	0.04	0.04	0.11	0.07	0.11	0.11	0.00	0.07	0.11	0.04
7	0.11	0.08	0.08	0.11	0.08	0.00	0.00	0.03	0.08	0.00	0.11	0.05	0.03	0.11	0.11
8	0.02	0.11	0.09	0.04	0.07	0.03	0.05	0.10	0.13	0.09	0.09	0.02	0.06	0.06	0.03
9	0.09	0.05	0.10	0.07	0.11	0.02	0.06	0.10	0.11	0.04	0.03	0.06	0.05	0.08	0.03
10	0.05	0.08	0.08	0.08	0.06	0.06	0.05	0.09	0.13	0.06	0.05	0.06	0.06	0.03	0.05
11	0.04	0.09	0.09	0.08	0.12	0.03	0.08	0.05	0.04	0.05	0.12	0.08	0.03	0.09	0.01
12	0.09	0.05	0.09	0.13	0.09	0.00	0.03	0.08	0.13	0.06	0.03	0.09	0.05	0.02	0.05
13	0.05	0.07	0.09	0.05	0.09	0.07	0.05	0.17	0.03	0.00	0.12	0.03	0.14	0.05	0.00
14	0.00	0.03	0.14	0.03	0.16	0.03	0.00	0.17	0.12	0.07	0.07	0.07	0.02	0.03	0.05
15	0.00	0.06	0.09	0.06	0.15	0.00	0.03	0.12	0.12	0.03	0.06	0.09	0.06	0.15	0.00

4.2   Building Transition Probability Tables and Constructing Graphs

In the training step the artificial (SIFT) feature points which were found closest to the current eye position of the human gaze were chosen on each clustered image captured from the video clips. This process is a kind of fusion of the artificial and human vision measurements. The number of the switches between the clusters on each consecutive image pairs was counted, and tables were constructed as follows. For the image pairs, the cluster of the current image and the cluster of the next image were identified, and the number of the steps between these cluster-pair was counted. For our purpose, transition probability tables were constructed about the given videos by normalizing the transition tables (Table 1).

According to the tables, the so-called transition probability graphs were constructed using the possible artificial feature points and the related cluster data of the transition probability tables in order to model the eye movement. Two different types of graphs were generated, which have some common properties as well. The connections between the vertices and the weight of the edges were computed from the transition probability tables. The vertices of both kinds of graphs are the generated SIFT points. All of them were connected to some of the closest neighbor vertices, and the number of the connections was a determined

Fig. 3. Comparison of the two graphs construction approaches. The weights of the graphs are derived from the transitional probability tables. The images on the left show the graphs in which every vertex is connected to the 8 neighbor vertices. The images on the right show the graphs where the vertices are connected to the 3 neighboring vertices. The images in the upper row show the detected SIFT points as well.

parameter that was constant in the same graph. The weights of the edges were determined by normalizing the values of the transition probability table to 1.

In the first approach, the vertices of the graphs were connected to the N closest neighbor vertices. In our model, the optimal value of N was between 8 and 15. This is a rather complex graph in the sense that the most likely edges have a larger weight, while the less probably edges have a smaller one.

Another kind of graphs was constructed as well to simplify the graphs of the first approach (see the right images of Fig. 3). In this case, the number of connections between the vertices is lower. The simplification is based on statistical solutions where the connected vertices are assigned within a given radius. The edges with the maximal weights are located about the same place in the two kind of graphs.

4.3   Estimation of the Most Probable Points - Regions of Interest

The goal of this step is to find the most attractive saliency regions of the human fixation pattern.

In the graphs, an inverse approach was applied to determine the most possible attention points: the probability of arriving at the given point was determined by this statistical summation. The values of the possibility were then summed and normalized to 1. As a result, a statistical probability was calculated, where only the points above a certain probability were taken into account and were marked as interesting points. Points below a given threshold were filtered out.

The left side of Fig. 4 shows some estimation results. The most attractive points inferred by our model are illustrated by purple dots. However, these results can be improved by taking behavior during the perception phase into consideration. This methodology is presented in Subsection 5.3.

5   Verification

5.1   Psychophysical Evaluation Experiments by a Desktop Eye-Tracker

The results of the estimation were evaluated by human test experiments in which we strictly followed the instructions of psychophysical experts. Ten volunteers were asked to take part in at least two different experiments one by one. Images were displayed to volunteers in a large resolution screen (1280x1024), their eye movement was recorded by another eye-tracker [3] and annotated to the images for analysis. This device was set on the desktop in the front of the monitor within a 50 cm distance. The subjects had to be motionless during the experiment.

According to some studies [8], different tasks can cause different behavior during human vision. Therefore, different tasks were given to the subjects: in some cases eye movement was recorded during free patrol, in other cases they

Fig. 4. Comparison of the most probable gaze points to a real human gaze path of one observer recorded during the evaluation experiments. The left image shows the estimated points based on the transition probability graphs with statistical summary, which were trained by evaluating training videos. For subjective evaluation, the right image shows a typical human eye movement on the same image from one observer. The fading color intensity shows the temporal change of the eye movement.

had to memorize the images, or search for some strange parts on the images. The different recorded eye movement patterns show that the eye movement is task-dependent. The displayed images are derived from our collection or from public databases [4], [5]. Between the images, there were some seconds to relax. Before displaying the images, a fixation cross was presented, therefore the position of the gaze was set to the center of the image.

The results of the test experiments were transformed into a fixation density map. This kind of visual attention map shows, how attractive are the parts of the image in a visual sense, therefore it shows whether a point of the image is a part of the fixation region. A fixation map containing only the fixation regions was filtered by a Gaussian distribution function. The density value was calculated for all the pixels of the images by

$$s(x, y) = \frac{1}{N} \sum_{n=1}^{N} \frac{1}{2\pi\sigma_s^2} \exp\left(-\frac{(x - x_n)^2 + (y - y_n)^2}{2\sigma_s^2}\right) \tag{1}$$

where (x_n, y_n) is the location of nth fixation point and σ_s^2 is the standard deviation of the Gaussian distribution. Fig. 5 presents a test image and its fixation density map.

5.2   Detection of the Fixation Regions on Human Experiments

Human fixation patterns from the test experiments were superimposed on each image. The right image of Fig. 4 is an example for the path of the human subjects. There are different adaptive algorithms to identify the different eye

Fig. 5. An input image with human experiment-based regions of interest and its human fixation density map. After the test experiments, the fixation regions were calculated based on the different speed of eye movement. The labeled images were then converted to the fixation density map. On this map, every pixel is compared to the entire fixation points, following the formula explained in the text. As a result, this kind of two-color map is generated where the color intensity is proportional to local image saliency.

movement patterns, e.g. [14], where identification of the fixation points is based on the velocity-difference between the fixation and saccade points.

The desktop eye tracker has some simpler but appropriate software for this task, it is possible to distinguish between the two types of points, differentiating them by a given velocity threshold that can be parameterized. The method is based on the distance between the neighboring points, assuming the temporal constancy of the sampling. However, we assumed that the movement of the eye consists of only fixation and saccade points, and we did not pay attention to the glissade (micro-saccade) periods like the software environment of the wearable eye-tracker. This way, groups of fixation points were collected and painted to the original images as well.

5.3   Filtering the Estimated Fixation Patterns

Since filtering the estimated fixation patterns influences the results, we present this selection method in this section. The goal of filtering the estimated points is to improve the reliability of our model, therefore; we statistically analyzed the biology-driven behavior of the users detecting statistical correlation.

In the training videos, the movement and the current position of the head was taken into account. We can assume that the head usually turns into the direction of activity, objects and interesting landmarks. As a result, the most attractive points are usually in the center of the images.

Although bottom-up properties of the peripheral regions can be similar to the points located at the center of the images, their importance is less for the subjects due to some top-down factors. In other words, the probability that a peripheral region is attractive for a human subject is less likely, the probability of arriving to that point by the gaze can be reduced. The extent of the distance-based filtering was determined by the analysis of the human test experiments. The images were divided into rings which had the same origo, while their radius was different. The spatial distribution of the detected fixation points was analyzed and behavior-tables were created including transitional probability tables about the movements from one fixation point to the next one. After generalization, the numerical results were integrated to our model, and probabilies of the peripheral edges were reduced according to their distance to the center.

5.4   Comparison of the Estimated Points to Real Measurements

To compare the prediction of our model, the spatial saliency map of the images were created. Similarly to the fixation density maps of the test experiments, the spatial saliency map shows how attractive an image region of the image is.

For the objective comparison of the artificially inferred points to other state-of-the-art methods ([11] and [10]) and to human test experiments, the symmetric Kullback-Leibler distance was used. After normailizing the saliency maps to one, they can be considered as two dimensional probability density functions. In the unattractive regions (black parts of the saliency maps) the value of the function is actually zero, while in the other parts the function is less than 1.

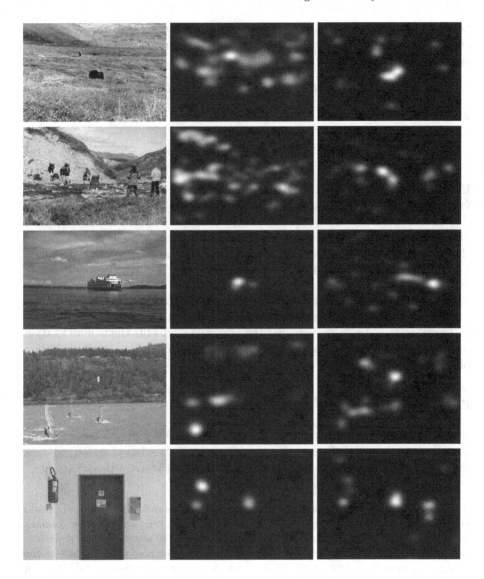

Fig. 6. Comparison of the artificial fixation patterns on Images #1-5 to the real human salient regions. The input images are in the first column, the second column contains the artificial saliency maps generated by our model. The color intensity of white shows how a local region was labeled as salient by applying our approach. Human fixation density maps can be found in the third column, these maps were constructed by summing the fixation density maps of each observer. Besides these maps show the salient regions of the images, they are needed for the objective comparison of the human and artificial regions of interest by calculating the Kullback-Leibler distance.

Let us denote the probability function of the human measurements by $P(x)$ and the probability function of the artificially measured points by $Q(x)$. The symmetric Kullback-Leibler distance is

$$D_{KL}(P,Q) = D_{KL}(P||Q) + D_{KL}(Q||P) \tag{2}$$

where

$$D_{KL}(P||Q) = \sum_i P(i)log\left(\frac{P(i)}{Q(i)}\right). \tag{3}$$

If the two density function is the same, the KLD value is 0. If the division in the formula would be zero, for that i value the sum is zero as well.

The estimated points of each test experiment were transformed into fixation density maps, and these maps were summed into one map. The comparison of the artificial saliency maps and the human fixation density maps can be found on Fig. 6. The points determined by our model describe the most probable points of the path that a human subject patrolled with his or her eyes. Our data set consists of 50 images. The numerical results of our data set can be found on Fig. 7.a. Some examples are presentend on Table 2 and on Fig. 7.b.

These numerical results show that the proposed model fits the human fixation measurements better than any previous method. Our model is a good prediction of human-oriented saliency areas on images and videos.

6   Conclusion

In this paper, we have presented an algorithm for inferring the eye movement on images and even on real scene situations for which HVS-based statistical data were used discovered in our experiments and in recent state-of-the-art studies.

Table 2. The numerical results. The first column contains the KLD distance between our prediction and the averaged human experiments. The second and the third columns show the similar results of two state-of-the-art methods. The averaged fixation density map was compared to the uniform distribution as a reference: the accuracy of the model reduces if the transition probabilty table is a uniform distribution table (fourth column). The last column is a reference where the fixation density maps have uniform distribution.

	Our model vs. human	Itti vs. human	GBVS vs. human	Random graphs vs. human	Uniform distribution vs. human
# 1	1.524	1.222	1.099	1.806	1.863
# 2	1.162	1.138	1.169	1.412	1.641
# 3	0.251	0.958	0.897	0.915	1.524
# 4	0.634	1.134	0.968	1.310	1.465
# 5	0.344	1.749	1.657	2.092	2.391

Our proposed method can give a good estimation of visual saliency without object detection. It is based on low-level features (artificial saliency points) which are connected in a graph based on the transitional probabilities among feature points measured in human experiments. Moreover, the proposed method may give a good estimation of possible eye paths through the image as well; artificial saliency points where the graph of possible eye paths is dense may assign region of interest areas.

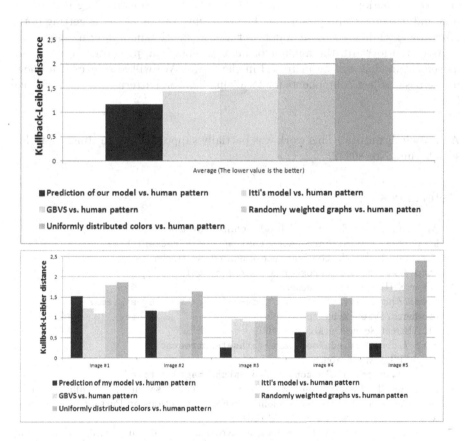

Fig. 7. Diagrams about the results of the predictions. Our data set contains 50 images. In both cases, the vertical axis is the Kullback-Leibler distance between the saliency maps of the algorithmic predictions and the human fixation density patterns. The lower this value is, the more accurate the prediction is. The first column contains the KLD distance between our model and the human patterns. The fourth column shows the distance, if the transition probability tables have uniform distribution. This means that biological data does significantly influence the results. On the second diagram, the horizontal axis shows some images taken from the test experiments. The last column shows a reference where the fixation density maps have uniform distribution. Our results are compared to two state-of-the-art methods: [10] and [11].

In the model, we combined two approaches. During the video records, some kind of top-down features were recorded and used as a training set. During the processing of the captured frames, bottom-up methodology was applied taken from image processing algorithms.

The paper describes an HVS-based statistic as well in order to improve the model of the human eye movement and to estimate visual pathways on images and scenes. Statistical data about the correlation of the eye and assumptions about head movement were also integrated and combined to a training-based graph-construction algorithm which is a new approach for modeling human vision and it seems to be promising. The reliability of our model is getting optimized by extending the human data set and by developing a filtering model.

As a step forward, the amount of built-in biological, perceptual and head movement models will be increased in the future. We will also investigate how to use basic object components for localizing fixation patterns instead of simple feature points.

Acknowledgments. This work was partially supported by the Hungarian Research Fund No. 80352.

References

1. Arrington Research, GigE-60 head-mounted eye-tracker,
   http://www.arringtonresearch.com/LaptopEyeTracker1.html
2. Arrington Research, GigE-60 head-mounted eye-tracker (image),
   http://www.arringtonresearch.com/news.html
3. iView X Hi-Speed eye-tracker,
   http://www.smivision.com/en/gaze-and-eye-tracking-systems/
   products/iview-x-hi-speed.htm
4. MIT Street Scene Database (2007),
   http://cbcl.mit.edu/software-datasets/streetscenes/
5. Ground Truth Database (2009),
   http://www.cs.washington.edu/research/imagedatabase/
6. Avraham, T., Lindenbaum, M.: Esaliency (extended saliency): Meaningful attention using stochastic image modeling. IEEE Transactions on Pattern Analysis and Machine Intelligence 32, 693–708 (2010)
7. Banko, E.M., Gal, V., Kortvelyes, J., Kovacs, G., Vidnyanszky, Z.: Dissociating the effect of noise on sensory processing and overall decision difficulty. Journal of Neuroscience 31(7), 2663–2674 (2011)
8. Betz, T., Kietzmann, T.C., Wilming, N., Konig, P.: Investigating task-dependent top-down effects on overt visual attention. Journal of vision 10(3) (2010)
9. Csurka, G., Dance, C., Fan, L., Willamowski, J., Bray, C.: Visual categorization with bags of keypoints (2004)
10. Harel, J., Koch, C., Perona, P.: Graph-based visual saliency. In: Advances in Neural Information Processing Systems 19, pp. 545–552. MIT Press (2007)
11. Itti, L., Koch, C., Niebur, E.: A model of saliency-based visual attention for rapid scene analysis. IEEE Trans. Pattern Analysis and Machine Intelligence 20(11), 1254–1259 (1998)

12. Liu, T., Yuan, Z., Sun, J., Wang, J., Zheng, N., Tang, X., Shum, H.Y.: Learning to detect a salient object. IEEE Transactions on Pattern Analysis and Machine Intelligence 33, 353–367 (2011)
13. Lowe, D.G.: Distinctive Image Features from Scale-Invariant Keypoints. International Journal of Computer Vision 60, 91–110 (2004)
14. Nyström, M., Holmqvist, K.: An adaptive algorithm for fixation, saccade, and glissade detection in eyetracking data. Behavior Research Methods 42, 188–204 (2010)
15. Sivic, J., Zisserman, A.: Video google: a text retrieval approach to object matching in videos. In: Ninth IEEE International Conference on Computer Vision, vol. 2, pp. 1470–1477 (2003)
16. Smith, K., Ba, S., Odobez, J.M., Gatica-Perez, D.: Tracking the Visual Focus of Attention for a Varying Number of Wandering People. IEEE Transactions on Pattern Analysis and Machine Intelligence 30(7), 1212–1229 (2008)
17. Szolgay, D., Benedek, C., Sziranyi, T.: Fast template matching for measuring visit frequencies of dynamic web advertisements. In: International Conference on Computer Vision Theory and Applications (VISAPP), Porto, Portugal (2008)
18. Tseng, P.H.H., Carmi, R., Cameron, I.G., Munoz, D.P., Itti, L.: Quantifying center bias of observers in free viewing of dynamic natural scenes. Journal of vision 9(7) (2009)

Author Index